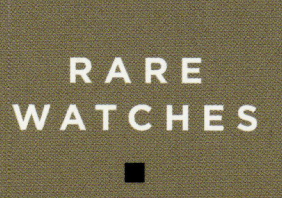

RARE
WATCHES

Paul Miquel

First published in France in 2018 by Gründ
Concept by Olo Éditions

First published in Great Britain in 2019 by Conran Octopus Ltd,
an imprint of Octopus Publishing Group Ltd
Carmelite House
50 Victoria Embankment
London EC4Y 0DZ
www.octopusbooks.co.uk

An Hachette UK Company
www.hachette.co.uk

The authorized representative in the EEA is Hachette Ireland, 8 Castlecourt Centre,
Dublin 15, D15 XTP3, Ireland (email: info@hbgi.ie)

Distributed in the US by Hachette Book Group
1290 Avenue of the Americas
4th and 5th Floors
New York, NY 10104

Distributed in Canada by Canadian Manda Group
664 Annette St.
Toronto, Ontario, Canada M6S 2C8

ISBN 978-1-84091-783-3

A CIP catalogue record for this book is available from the British Library.

Printed and bound in China.

20 19 18 17 16 15 14

Watch dimensions are given in millimeters to ensure accuracy, as is industry standard;
other measurements are given in both imperial and metric.

For this edition:
Senior Commissioning Editor: Joe Cottington
English Translator: Simon Jones
Assistant Editors: Emily Brickell and Katie Lumsden
Designer: Jack Storey

For Olo Éditions
Publisher: Claudie Souchet (assisted by Jeanne Daucé)
Picture Researcher: Jeanne Daucé
Creative Director: Philippe Marchand
Art Director: Émilie Greenberg
Graphic Designers: Nicolas Marchand and Thomas Hamel
Photo Reproduction: Peggy Huynh-Quan-Suu
Production: Stéphanie Parlange

I have had the privilege of being surrounded by rare watches for three decades, more than two of those in a professional capacity. It would be understandable if, after so many years, even the most extraordinary watches had eventually become the norm, if I had lost the discernment, the emotion, the sense of their rarity. Yet I have been lucky in being able to retain this emotion intact – to experience the same thrill every time I set eyes on an unusual model with a singular history.

For some, a watch's value is measured in terms of its absolute rarity – that is, the number of known pieces. For others, it is measured by its price to the public, or the sale price it realizes at auction. For others still, the value relates to the watch's mechanical complexity, its history or simply its beauty. For most of us, it is all these factors together that make us appreciate a rare watch.

For my part, the best indicator of value is how fast my heart beats when I hold a watch in the palm of my hand. Emotion is the real driver, whether it comes from a watch's design, manufacture and sale (for the professional), or from seeking out, discovering, admiring and possessing a watch (for the private collector). I am fortunate that I experience these emotions doubly – as both a professional enthusiast and an enthusiastic professional.

In the course of my career I have worked for various prestigious auction houses, where I have been able to come across, and sell at auction, a large number of exceptional watches. I am delighted to find, in the first chapter of this book, several museum pieces that I had the honour of coming into contact with at some point in their existence. I still remember the emotion I felt when I held each of these pieces for the first time, then studied them in preparation for sale, and right up to when they went under the hammer – always going for a record price. Those few months of preparation were always full of thrills, fascinating discoveries and surprises.

I hope, dear reader, that when perusing these pages you will feel the same delight a collector experiences on discovering a rare watch.

AUREL
BACS

A mechanical watch made to the highest standards, whether in the 17th century or the 21st, will still be reparable after 100 years – or 1,000 years, or even longer. This durability, which sets the art of watchmaking apart, can never be equalled by electronic technology or computers, which, marvellous though they are, will be condemned sooner or later to obsolescence. The art of traditional mechanical watchmaking, by its very nature, bears the stamp of eternity.

Paul Miquel's book casts a new and original light on the multiple facets of the watchmaker's art. The pertinence of his choices and notes demonstrate the author's expertise and passion. Far from being a demanding academic work, this book can be read and referred to as a fairytale, bringing the enchantment and emotional power of this art within everybody's reach.

As timepieces of yesterday, today and tomorrow, mechanical watches bear moving witness to the genius and skill of humans, whom in turn they have accompanied throughout the centuries. This is what gives the watchmaker's art its power to move, to transmit knowledge and to last for eternity.

JEAN-CLAUDE BIVER

■

STARS AT AUCTION

ROLEX COSMOGRAPH DAYTONA

REFERENCE 6239 'PAUL NEWMAN'

■

THE HOLY GRAIL

A benchmark among vintage chronographs, the 'Paul Newman' Rolex watches (references 6239, 6241, 6262, 6263, 6264 and 6265) became icons of watchmaking in the space of just a few years. The American actor's Rolex watch beat all sales records on Thursday 26 October 2017 when it was sold at auction for $17.75 million (£13.83 million) by the auctioneer Aurel Bacs, at Phillips in New York. Here is the true story of the world's most expensive wristwatch: Paul Newman's Rolex Daytona reference 6239, an object of veneration.

The year was 1965. Paul Newman, a gifted pilot and a speed junkie, was involved in a motorcycle accident. He survived; but for his wife, American actress Joanne Woodward, it was the final straw. It is said she could no longer tolerate being worried sick every time her daredevil of a husband lined up at the start of a car race or recklessly rode his Triumph motorbike. Legend has it that no matter how much she exhorted him to be sensible, he wouldn't listen. In 1968, in New York, she dashed to Tiffany & Co. and bought a sports watch, a Rolex Cosmograph Daytona, reference 6239 in steel, with a dial known as 'exotic'. It had been created in 1962 and was made for motor racing, with a tachymeter scale engraved on its bezel. Woodward had it engraved with the words 'DRIVE CAREFULLY ME'. She kept it hidden in a drawer for a few months and finally gave it to him in 1969, when he was shooting the film *Winning*, directed by James Goldstone, in which Newman played racing driver Frank Capua and Woodward played his wife. Newman wore his new watch during the movie, and immediately fell in love with it. He boasted of its precision to his friends, and even organized competitions, phoning the speaking clock to demonstrate its incredible accuracy.

In 1962 Rolex was made the official timekeeper of the Daytona 500 and designed this watch in honour of the legendary race. To this day, winners of the Daytona 500 are presented with one. After the shoot of *Winning*, Paul Newman became superstitious. In town and on the track, because he loved it and had good taste, he wore his Daytona watch every day, as numerous photographs of him confirm. During the 1980s, collectors nicknamed these chronographs 'Paul Newman'. So the legend was born. 'The models with exotic dials, which are highly sought-after, soon became associated with Paul Newman, the actor and racing driver,' Rolex explains. 'No one knows exactly how his name came to be linked to these watches, but the Cosmograph Daytona certainly

SOLD FOR $17.75 MILLION (£13.83 MILLION) – A WORLD RECORD

gained in popularity when Paul Newman wore it on the big screen in the movie *Winning*. From then on, aficionados referred to this watch as the "Paul Newman Daytona", even though Rolex had never used that label.' Since then, not only the Rolex Daytona reference 6239, but also references 6241, 6262, 6263, 6264 and 6265 have been known as 'Paul Newman'. What are their specific features? They have 'panda' dials, with square markers on the subdials to make it easier to read the divisions of time.

In the mid-1980s Newman's watch disappeared from view. No one knew what had become of it. What people did know was that Joanne Woodward had given her husband another one – a reference 6263. It was believed that the original 6239 was lost. Wrong. In the summer of 1984, Elinor ('Nell'), the actor's daughter, and her boyfriend at the time, James Cox, were staying at the Newmans' house. James Cox, a DIY enthusiast, was helping Paul Newman rebuild a cabin among the trees in the garden. 'Pop would frequent the riverbank to check on James' progress,' Elinor later recalled. 'During one such encounter, Pop asked James if he knew the time. Apparently Pop forgot to wind his wristwatch that morning. James responded that he didn't know the time and didn't own a watch. Pop handed James his Rolex and said, "If you can remember to wind this each day, it tells pretty good time".' From then on, James Cox meticulously looked after the watch, until it reappeared as if by magic in October 2017 at an auction at Phillips in New York, held in collaboration with Bacs & Russo, which offered 50 exceptional watches for sale. It took just 12 minutes, before 700 people, for the real 'Paul Newman' Daytona to smash all expectations and sell for $17.75 million (£13.83 million) – a world record – to an anonymous buyer. There have been rumours that this buyer does not, in fact, exist and that the watch was bought by a group of collectors with the unavowed aim of boosting the increasingly unrealistic price of sporting Rolex watches. Myth or reality? The secret remains a closely guarded one.

PATEK PHILIPPE HENRY GRAVES JR

SUPERCOMPLICATION, 1933

■

THE IDOL OF GRAND COMPLICATIONS

Patek Philippe's 'Henry Graves Jr Supercomplication' is the watch that brings together all the superlatives. With 24 complications, it was the world's most complex watch for 56 years. In November 2014 this 1933 pocket watch with two dials, made entirely by hand, sold for $24 m (£18.7 million) at Sotheby's in Geneva. This unusual piece was certainly a masterpiece of watchma but its incredible story and that of its owner, Henry Graves Jr, make it a true legend.

In the early 20th century, in the USA, two men decided to compete against each other in an unusual field - watchmaking. One was James Ward Packard, a big motorcar manufacturer from Ohio; the other was Henry Graves Jr, a banker with a passion for sport and art. Both were rich, powerful and just a little megalomaniacal. They were customers of Patek Philippe and loved special orders. In January 1916 James Ward Packard received an exceptional pocket watch, with 16 complications; in April 1927 he added another pocket watch to his collection, featuring 10 grand complications, including a celestial chart. But James Ward Packard's timepieces, exceptional though they were, were far from being the most complex ever developed; it was difficult to compete with masterpieces of watchmaking complexity such as the Leroy 01 and Breguet's 'Marie-Antoinette'. Packard and Graves loved complicated watches and competition. Their goal was simple: to own the world's most complicated watch. In 1925, with the utmost secrecy, Henry Graves Jr placed a very special order with the watchmakers at Patek Philippe. The spec sheet could be summed up in a single line: he wanted, quite simply, the world's most complicated watch. It didn't matter how long it would take, how much it would cost or what complications it would feature – that was not the point. The watchmakers at Patek Philippe agreed to take on the challenge. 'Just' eight years would be enough to accomplish this huge task. In complete secrecy, the Patek Philippe pocket watch no. 198 385 was delivered to its buyer on 19 January 1933 – meaning that Henry Graves Jr beat James Ward Packard hands down. It was a pocket chronograph in gold, with a double face, minute-repeater and Westminster chimes with carillon.

Here are a few numbers to give an idea of this masterpiece, which cost the astronomical sum of 60,000 Swiss francs: 535 grams, 920 individual parts including 430 screws, 110 wheels, 70 rubies and 120 levers, springs, cams and clutch levers, for a total of 24 complications.

The most exceptional of those 24 complications are: a perpetual calendar, the phases of the moon, sidereal time, a power reserve indicator, equation of time, and sunrise and sunset times for the city of New York. These exceptional features enabled Patek Philippe's 'Henry Graves Jr Supercomplication' to remain the world's most complex watch for half a century; it was dethroned in 1989 by another Patek Philippe creation, the 'Calibre 89', with 33 complications, which was produced for the company's 150th anniversary.

In order to personalize this masterpiece forever, 'Made for Henry Graves Jr' was inscribed on the dial of the 'Supercomplication', and his family's coat of arms and Latin motto ('*Esse quam videri*' – 'To be rather than to seem') were engraved on the solid yellow gold casing. This unique watch is the only example in yellow gold; its double, in platinum, is now part of the collection of the Patek Philippe Museum' in Geneva.

UST ONE WISH: THE WORLD'S MOST COMPLICATED WATCH

Between 1910 and 1953 Henry Graves Jr amassed the biggest private collection of Patek Philippe watches in the world. When he died in 1953, most of his collection was inherited by his daughter, Gwendolen, who in turn left it to her son, Reginald H 'Pete' Fullerton (1933–2012) – himself a collector – in 1960. In 1969 Fullerton sold the 'Henry Graves Jr Supercomplication' to Seth G Atwood, a respected collector of clocks and watches, who founded the Time Museum in Rockford, Illinois. The watch was put up for auction for the first time in 1999, during a sale organized by Sotheby's that offered 81 exceptional pieces from the Time Museum. The watch was estimated at $3–5 million (£2.3–3.9 million), but it exceeded all expectations: a private collector paid $11 million (£8.6 million) – the highest price ever paid for a watch at that time – to secure this legend of the watchmaker's craft.

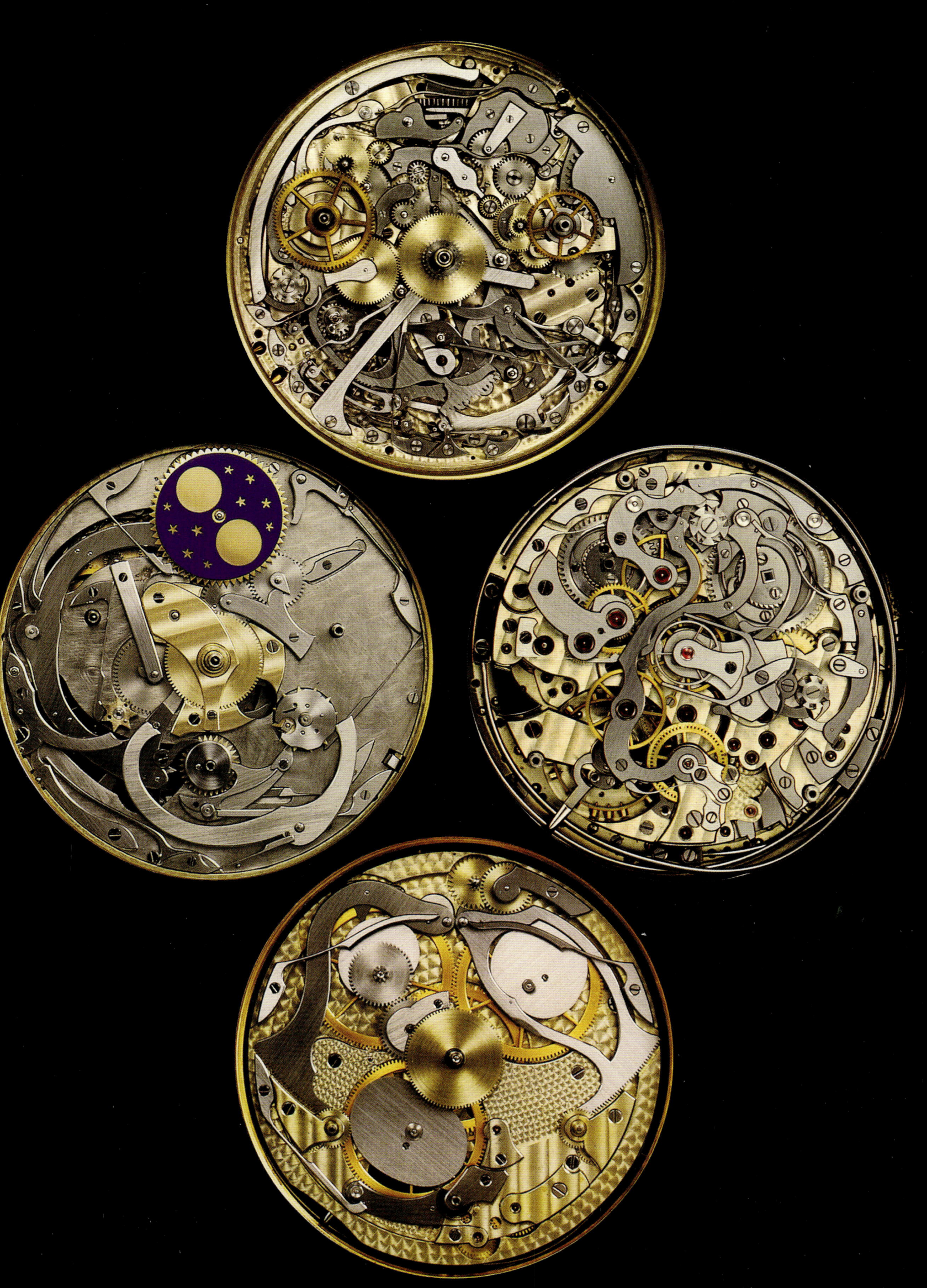

Fifteen years later, on 11 November 2014, this beauty resurfaced at another Sotheby's auction, held in Geneva. And there, time stopped. 'The list of superlatives which can be attached to this icon of the 20th century is truly extraordinary. Indisputably the "Holy Grail" of watches, the "Henry Graves Supercomplication" combines the Renaissance ideal of the unity of beauty and craftsmanship with the apogee of science,' Sotheby's stated at the time. 'Our offering of this horological work of art in 1999 was unquestionably the highlight of our professional careers and set a world record which has held until today. We are extremely privileged to be offering it once again.' At the time, the highest estimates were touching on £11.5 million ($14.8 million) – the wildest extravagance.

'THE LIST OF SUPERLATIVES WHICH CAN BE ATTACHED TO THIS ICON OF THE 20TH CENTURY IS TRULY EXTRAORDINARY'

'For some time I have been thinking about the value of this treasure,' explained the expert Aurel Bacs, who worked for Christie's at the time, in his article for the Worldtempus website. 'Over and over, I remind myself that it is, thanks to its uniqueness, history, number of complications and, not to forget, its distinguished signature on the dial, the most valuable watch in private hands today. Full stop.' The sale took place in a fevered, totally surreal atmosphere, in the immediate aftermath of Patek Philippe's 175th anniversary. After a bidding war that lasted a quarter of an hour, the 'Henry Graves Jr Supercomplication' sold for £17,799,500 ($22,854,600) – a record for a watch that, according to Aurel Bacs, 'is now officially sitting on top of the Mount Olympus of the art world as one of the most important and valuable works of art ever created by mankind, amongst the historically most relevant Impressionist paintings, the finest Ming-dynasty vases, the most fabulous motorcars and fascinating diamonds.'

ROLEX OYSTER PERPETUAL, REFERENCE 5029/5028.

ROLEX 'TWO AMERICAS', REFERENCE 6284.

ROLEX REFERENCE 8382

■

THE ELEGANCE OF ROLEX'S CLOISONNÉ WATCHES

The Rolex watches with cloisonné enamel faces created by the famous enamellist Marguerite Koch during the 1950s are extremely rare, legendary pieces that all collectors dream of. A unique 1949 model made headlines when it sold for more than £750,000 ($950,000) at a Christie's auction held in Geneva in May 2014. Its dial features a whale and a frigate adrift in a wild sea.

While her name may be totally unknown to the general public, in the small world of watch collectors Marguerite Koch is a true legend. During the 1940s and 1950s, this artist from Geneva made a very small number of watch dials in cloisonné enamel, chiefly for Patek Philippe and Vacheron Constantin. She was a highly gifted enamellist and was employed by the Stern Frères company. She made her mark, notably with designs depicting sailing ships and coats of arms for Rolex, and her masterly creation of astonishing dials inspired by the exotic pictorial world of Paul Gauguin, such as a 'virgin forest' Patek Philippe reference 2481, dating from 1953. Today these watches attract the greatest admiration, not just for their rarity, but for the method used to decorate their dials; cloisonné enamel is an art in its own right. The decoration of watch dials, which first appeared during the Renaissance, can require the use of a number of artistic techniques, of which enamel is one of the most impressive. In watchmaking, as in jewellery, two principal enamelwork processes are used together, *champlevé* enamelling and cloisonné.

To create *champlevé* enamel, the artist first chisels out hollows that are then filled with a small quantity of coloured enamel; each colour then requires firing in a kiln. The technique of cloisonné enamel, used to make unusual jewellery and certain Chinese vases, requires the greatest finesse. The outlines of the design are marked using fine metal ribbon or gold thread, which are secured to the surface or background. The enamellist generally works from a large-format image – a painting or photograph – before producing a design of the desired scale, either on the piece itself or on another medium of the same size. The aim then is to colour the partitions marked by the outlines.

Previous page: 1953 Rolex reference 8382 in yellow gold. Its dial, made by Stern Frères and partly covered in coloured cloisonné enamel, is inspired by Roman mythology. It depicts Neptune, god of freshwater and the sea, and protector of fishermen, boatmen and horses. It sold for more than £500,000 ($640,000) in May 2011 at a Christie's auction in Geneva.

Opposite: 1949 Rolex Oyster Perpetual, reference 5029/5028.

A multicoloured watch face with a complex design might require up to 70 applications of enamel and 15 firings in a kiln. The enamelled plate must then be smoothed or polished to bring out the delicacy of the gold ribbons and the brightness of the colours. This is the technique that Rolex chose for its enamelled watch dials.

The rare Marguerite Koch pieces that are still on the market include one unique masterpiece. Its dial is completely decorated with cloisonné enamel – a depiction of a nautical scene in which a majestic frigate and a whale make light of the elements in a raging sea. This Rolex Oyster Perpetual (automatic movement) watch, reference 5029/5028, dates from 1949 and has a gold casing with a diameter of 36mm (very impressive for the time). Many of its tiny details would make the most jaded collector

MARGUERITE KOCH, EXPERT IN THE NOBLE ART OF CLOISONNÉ ENAMEL

swoon. This extremely rare piece went on sale at a Christie's auction in Geneva on 12 May 2014, where it sold for £840,300 ($1,079,000). It is one of the most expensive Rolex watches in the world and, according to the expert's notes supplied by Christie's, 'can arguably be considered one of the most important Rolex watches ever manufactured'. There are several reasons for this: the dial is enamelled over its entire surface, not just over two-thirds of it as is more typical; the figures are replaced by stars; the brand logo (a crown) is at 6 o'clock and protected by the enamel; and the decoration, which is of breathtaking quality, features powerful colours and teems with tiny details that can be seen only under a microscope.

Left: On 27 March 2011 in Geneva, the Rolex 'Two Americas', reference 6284, was snapped up for £507,500 ($652,000), a record at the time for this type of Rolex. Its casing is 14-carat yellow gold and features an automatic movement. Only six were ever made.

BREGUET ET FILS

POCKET WATCH WITH TWO MOVEMENTS, BREGUET NO. 2667, 1814

■

THE BEAUTY OF THE FIRST CHRONOMETERS

One of three Breguet reference no. 2667 watches dating from 1814 was bought by the Breguet Museum and its current CEO, Marc A Hayek, for more than £3 million ($3.8 million) at a Christie's auction held in Geneva on 14 May 2012. This extremely rare flat pocket watch with two movements, based on the principle of the resonance chronometer, is, as of today, the most expensive vintage Breguet watch sold at auction.

In 1814 Abraham-Louis Breguet, a true watchmaking genius, developed a revolutionary pocket watch. It was very flat, featured two movements and was based on the principle of resonance chronometers. It was inspired by the work of the French watchmaker Antide Janvier (1751–1835), who noted that two oscillating bodies close together influenced each other reciprocally. 'This sophisticated technique implies two fully independent movements: two mainsprings, two geartrains, two lever escapements and two balances,' explains the Fondation de la Haute Horlogerie. 'A rack and pinion adjusts the distance between the two balances (and their escapement). Through such precise adjustment a position is found where, through the effect of resonance, the balances continually correct each other to maintain maximum precision.'

Breguet made three pieces using this technique. It is the first of the three, no. 2667, that we are dealing with here: a piece in 18-carat yellow gold, 63.7mm in diameter. It has two different dials: the first shows hours in Arabic numerals, the second in Roman numerals. According to a statement from Breguet, which has been part of the Swatch Group since 1999, this amazing timepiece simultaneously symbolizes 'the aesthetic appeal of Breguet, but also its inventive genius'.

Experimental in conception and conventional in form, this watch was sold for 5,000 francs in August 1814, in London, to a gentleman named Mr Garcias. It was returned to Paris on an unknown date, through the agency of someone called Gabriel, and was then sold for 4,500 francs on 30 April 1856 to Eugène Emmanuel of Savoy-Villafranca-Soissons (1816–88), Prince of Carignan and Count of Villafranca.

BREGUET EQUATION OF TIME WATCH, NO. 4111, 1827

■

WHEN BREGUET MADE LIGHT OF THE EQUATION OF TIME

Acquired by the Breguet Museum for nearly £2 million ($2.5 million) at Christie's in Geneva in May 2012, reference no. 4111 is a marvel. This Grand Complication in gold and silver, dating from 1827, is a concentration of all the genius of Breguet. It was the second vintage Breguet watch to have been the most expensive ever bought in the company's history at the time of its purchase.

Watches that feature the equation of time display the difference between 'true' solar time – that of nature – and the 'mean' solar time of our human societies. 'As it goes round the sun, the earth follows an elliptical trajectory. Moreover, its axis is tilted in relation to the plane of the equator,' explains the dictionary of the Fondation de la Haute Horlogerie. 'The "true" day – that is, the period of time between two "true" noons (when the sun passes its highest point in the sky) is therefore not the same length throughout the year. It lasts exactly 24 hours only four times a year: on 15 April, 14 June, 1 September and 24 December. This difference – which ranges from minus 16 minutes 23 seconds on 4 November to plus 14 minutes 22 seconds on 11 February – is called the equation of time. And watchmakers have always tried to outdo each other in using their ingenuity to reproduce this celestial mechanism.' Between 1790 and 1830 Abraham-Louis Breguet and his son Antoine-Louis manufactured 15 watches with the equation of time, some of which displayed the difference between true solar time and mean solar time via a small window that enabled the owner to increase or reduce the difference, from minus 16 minutes to plus 14 minutes. Others were known as *équation marchante* (moving equation) and featured two minute hands: one for solar time and the other for mean time. Reference no. 4111 is yet another type; it is one of a series of five watches featuring the equation of time that have two independent dials, allowing optimal reading of solar time and real time. Two of these five models also feature a minute-repeater mechanism that allows the hour, quarter of an hour and minutes displayed on the dial to chime 'on demand'.

Reference no. 4111 is one of these watches. This Grand Complication from 1827 is a flat watch in gold and silver with the equation of time and a repeater that chimes the hour, the half-hour, the quarter-hour and the half-quarter, 'on the principles of chronometers'. It features an annual calendar and a perpetual calendar. Mr Peyronnet, a Parisian banker, bought it for 7,500 francs on 10 January 1827; it was later taken back to Breguet, which sold it for 8,000 francs (certificate no. 1061) on 3 November 1834 to Count Charles de l'Espine. Then, 185 years after it was made, this pocket watch was one of the flagship lots at a legendary Christie's auction held in Geneva in May 2012, during which the House of Breguet bought back the two most expensive historic pieces in its history. One of them was the fabulous reference no. 4111, acquired by the Breguet Museum for £1,951,000 ($2,505,000).

■

THE MYSTERIOUS REFERENCE 4113

An aura of mystery surrounds Rolex reference 4113 watches. Only 12 units of this flyback chronograph, which has a history closely tied to motor racing, have been identified, and there are supposedly no more than eight on the market today. The appearance of one of them at an auction causes excitement amongst even the most jaded of collectors. On 14 May 2016 a record fell under the hammer of Aurel Bacs during a Phillips auction in Geneva: £1,838,400 ($2,360,500). Never before had such a sum been paid at auction for a Rolex watch. Here is the history of the '4113 chronos'.

S ome watches seem to possess supernatural powers; this may certainly be said of the 12 flyback chronographs, reference 4113, made by Rolex in 1942. It took years of research by experts and collectors to unravel the thread of their history. The reference 4113 is more than a legend: this chronograph is a unique case. Together with the 15 or so flyback chronographs ordered from Rolex by the Italian army in 1949, which bear the extremely rare reference 4110, the 4113 watches are the only chronographs made by Rolex that feature a flyback function (Valjoux 55), an oversized 44mm-diameter steel case and a very fine bezel, allowing the dial to be as large as possible. The chronograph is operated via three buttons: two rectangular ones on the edge of the case and a circular one integrated with the winding mechanism button. The dial is also highly readable. In the world of Rolex, these features are synonymous with extreme rarity. According to the best-informed collectors, only eight of these twelve pieces manufactured in 1942 are still in existence – and their appearances in public can be counted on the fingers of one hand.

The mystery surrounding these watches is sustained by Rolex, which made no announcement about them at the time they were made. Several experts, notably at Christie's, have investigated. They concluded that the reference 4113 chronographs are linked to the world of motor racing – specifically, to the Giro Automobilistico di Sicilia, which at the time was the longest road race in Europe. According to the auction catalogue for Christie's Geneva May 2011 sale, most of the Rolex reference 4113 watches that have reappeared on the market in recent years had a connection with Sicily. In fact, one of the very first to reappear – at a Christie's auction on 15 May 1991 in Geneva – belonged to the family of the famous Italian racing driver Stefano La Motta, Baron of Salinella, who wears it in several photographs. Another example – also sold by Christie's, this time in London, in October 1991 – was put up for sale by 'widow of a gentleman working for a racing team'. It may be that Rolex made these watches as rewards for the winners of certain motor races during the 1940s. The theory is a plausible one, but the mystery endures, arousing the interest of the greatest collectors. 'Without exaggeration,' notes Christie's, 'reference 4113 is broadly accepted to be the most sought-after and by far most valuable model in the universe of Rolex watches.'

VACHERON CONSTANTIN POCKET WATCH, A GIFT TO KING FUAD I OF EGYPT, 1929

■

WATCHMAKING AS DIPLOMACY

During the 1920s, when the eminent members of the Swiss diplomatic mission in Egypt decided to present a gift to King Fuad I, they naturally decided on a watch. The Egyptian monarch, a great watch enthusiast himself, appreciated the true value of the exceptional Vacheron Constantin grand complication pocket watch, given to him in November 1929, with its multitude of sophisticated complications. This historic watch sold for the astronomical sum of £2,532,500 ($3,252,000) at Antiquorum in 2005, on the 250th anniversary of Vacheron Constantin, the oldest Geneva manufacturer to have been continuously in business.

Quai des Moulins, Geneva, 1927. Francis Peter, a Swiss citizen and President of the Cairo Mixed Court, knocked discreetly on the door of the Vacheron Constantin workshops. He was on a mission: the small community of Swiss expatriates in Egypt had decided to make a gift of an exceptional watch to King Fuad I, who was planning an official visit to Switzerland. At that time, luxury watches were ordered directly from the watchmaker. So Francis Peter went to the design and manufacturing division of Vacheron Constantin that produced custom timepieces, which is today called l'Atelier Cabinotiers. 'At the time, Vacheron Constantin was working on an exceptional timepiece featuring a sophisticated collection of complications and eight hands,' relates Franco Cologni in the 2015 book *Vacheron Constantin, Artistes du Temps* (Flammarion). 'Once it was complete, it would be the most complicated watch the maker had ever created. It was this movement that won Francis Peter over. He decided to have the back of the timepiece decorated with the royal coat of arms, while the edge would be set with diamonds.' In October 1929 Francis Peter was finally able to take possession of the gift destined for the King of Egypt.

Being an adept diplomat, the Swiss magistrate requested some modifications. He knew the King was a Francophile, so asked that the display of the day and month be not in English but in French. A gift needs to be personalized, so the inside of the back cover, the 'bowl' of this very technically complex pocket watch, was engraved with a simple but sincere inscription: 'À Sa Majesté Fouad Ier, Hommage de la Colonie Suisse d'Égypte' ('To His Majesty Fuad I, with the respects of the Swiss colony of Egypt').

a key, in yellow gold, given as a gift by the Swiss authorities to the future King Farouk, son of King Fuad I, in 1935. The watch features a minute-repeater as well as a *grande sonnerie* and a *petite sonnerie* with three gongs struck by three hammers. It is equipped with a flyback chronograph with 30-minute totalizer, as well as a perpetual calendar, moon phase indicator, alarm, two power reserve indicators corresponding to the watch's train and the striking mechanism, with a movement comprising 16 rubies.

FUAD I, WITH THE RESPECTS OF THE SWISS COLONY OF EGYPT'

In November 1929 the timepiece was complete, meeting all the specifications requested by Francis Peter. It was presented to the monarch, the son of the Khedive Ismail, in a sandalwood casket adorned with an encrusted gold crown and the King's name inscribed in Arabic.

The watch, in yellow gold, features 13 complications, including a perpetual calendar, a flyback chronograph, a minute-repeater, a *grande sonnerie* and a *petite sonnerie*. One remarkable technical detail is that the watch features two barrels that are wound by the crown, the first for the hour and the second to supply the energy required for the chime. The dial is in solid silver and highly readable. Moreover, the back of the case is adorned with the royal coat of arms, an example of craftsmanship of rare refinement carried out by one of the most celebrated Geneva enamellists of the time. It is perhaps due to this remarkable watch that Farouk, Fuad I's son, also became a great watchmaking enthusiast. But that's another story.

PHILIPPE
GENEVE
ICELAND LONDON OSLO MOSCOW BAGHDAD
AZORES-MADEIRA PARIS GENEVA ISTANBUL
DAKAR ALGIERS ROMA CAPETOWN ADEN REUNION CEYLON
MAURITIUS ISLD
BOMBAY
CALCUTTA
SINGAPORE
SAÏGON
PEIPING TOKYO
ADELAÏDE
SYDNEY
RIO DE JANEIRO
BUENOS-AIRES
NEW-YORK
MONTRÉAL
CHICAGO
MEXICO
DENVER
CALIFORNIA
ALASKA
KLONDIKE
TAHITI

PATEK PHILIPPE WORLD TIME, REFERENCE 1415 IN PLATINUM, 1939

■

THE LOUIS COTTIER EFFECT

If one watch could symbolize the power Patek Philippe has over collectors, it would probably be this extremely rare 1939 World Time, reference 1415 in platinum, which sold for 6,603,500 Swiss Francs (over £2.8 million/$4 million) at a Geneva Antiquorum auction in the spring of 2002. What is the secret of the popularity of this two-century-old Swiss manufacturer?

'You never actually own a Patek Philippe. You merely look after it for the next generation.' In 1996 this Geneva manufacturer, founded in 1839, launched its new advertising campaign, dubbed 'Generations' and aimed at the general public. The slogan was chosen by the Stern family, who had owned the distinguished manufacturer since 1932, and it made an immediate impression: Patek Philippe watches were, above all, objects to be passed down from one generation to the next.

Those two sentences profoundly changed the brand's public image. Yet an advertising slogan alone cannot explain the all-powerful position occupied by Patek Philippe in the world of auctions. There are many other reasons for its supremacy: the extremely high quality of the watches the manufacturer has produced for 180 years, and the fact that Patek Philippe is still an independent, family business, with a historic heritage that is among the most prestigious. Today, we can only look with alarm at the stratospheric prices achieved by certain rare Patek Philippe watches in extraordinary auctions, an unquestionable sign of the appeal this brand has for collectors.

The utterly unreal price – 6,603,500 Swiss Francs – attained by the World Time reference 1415 at the April 2002 Antiquorum auction held in Geneva is probably the most striking example of the 'magic powers' of timepieces bearing the Patek Philippe name. Dating from 1939, this platinum watch, the only one of its kind, features a system for instantly telling the time in 41 cities, with their names engraved on its bezel. This 'small complication', called World Time, has always fascinated collectors, although it is not an example of high-level watchmaking. It was devised by Louis Cottier, an independent watchmaker who, at the beginning of the 1930s, invented this ingenious mechanism that makes it possible to tell the time in all time zones on a watch dial with a single glance.

In watchmaking circles this is known as the 'Cottier system': around the central dial, which has hands telling the local time, a disc displaying 24 hours automatically turns one notch anticlockwise every hour, while around it a fixed disc displays the cities in question. The first watch of this type ever made was a Vacheron Constantin – reference 3372, made in 1932 – featuring a dial bearing the names of 31 cities.

TELLING THE TIME IN 41 CITIES AT A GLANCE

The big names of watchmaking, including Patek Philippe, placed orders for this small jewel of ingenuity. Louis Cottier worked closely with Patek Philippe, to which he devoted a large proportion of his production: he would graft his mechanism, which continues to feature on many modern mechanical watches, onto unfinished pieces supplied by Patek Philippe. It was also Louis Cottier who created the very particular shape of the hands that are fitted to almost all vintage Patek Philippe World Time watches.

PATEK PHILIPPE
GENÈVE
ICELAND LONDON OSLO MOSCOW BAGHDAD BOMBAY SINGAPORE ADELAÏDE AUCKLAND ISLD FIJI ISLD ALASKA CALIFORNIA NEW-YORK RIO DE JANEIRO
AZORES MADEIRA PARIS GENEVA ISTANBUL MAURITIUS ISLD CALCUTTA PEIPING TOKYO SYDNEY MARSHALL ISLD SAMOA TAHITI KLONDIKE MEXICO CHICAGO BUENOS-AIRES
DAKAR ALGIERS ROMA CAPETOWN ADEN RÉUNION CEYLON SAÏGON DENVER MONTRÉAL

ROLEX OYSTER COSMOGRAPH DAYTONA 'ALBINO', REFERENCE 6263 IN STEEL, 1971

■

THE ROLEX THAT WENT OFF THE SCALE

The Rolex Cosmograph chronographs have become legends under the name 'Daytona', because of their links to the motor racing circuit in the city of Daytona Beach, Florida. These mythical pieces fascinate collectors. But the reference 6263, with its screw-down push buttons, has that extra something special that makes it stand out. Because the dial has a unique character, prices have soared: in May 2015 this 1971 Daytona, dubbed 'Albino' on account of its white dial, sold at Phillips for £1,015,000 ($1,303,300) – one of the most expensive Rolex watches ever sold.

This watch's destiny is certainly extraordinary. Made in 1971, the Rolex Cosmograph reference 6263 is much more than a high-precision mechanism for telling the time: it is a legend. Only four examples of this very rare model 'Daytona', with its 'Albino' dial, are known to exist. Usually, the colour of the subdials on reference 6263 watches – namely black – differs from that of the dial. For this non-standard Rolex, the subdials are the same colour as the dial – a fine silver. The monochrome nature of its dial, with its especially appealing gleam, makes this 1971 Daytona a highly sought-after rarity. Although many Daytona watches have undergone, over time, more or less conventional alterations, this one possesses an impeccable pedigree: a 727-calibre movement, original black bezel and MK1 screw-down push buttons.

The English guitarist Eric Clapton, a great watch collector, acquired it in the late 1990s. It later reappeared, first in New York on 5 June 2003 at a Christie's auction, alongside many other rare Rolex watches and some unique platinum Patek Philippe watches. It changed hands and was sold again a few years later at Sotheby's, for $505,000 (£393,400) – five times its pre-sale estimate. When it resurfaced, in May 2015 at Phillips in Geneva, experts knew that this chronograph with a singular destiny would go through the roof – but few could have foreseen that it would break all records and sell for almost $1.5 million (£1.17 million), becoming the most expensive Rolex ever sold.

This was an impressive performance for this simple steel chrono, equipped with a hand-wound chronograph movement based on a Valjoux 72 calibre. It has, however, a small quirk: the inside face of the back case is engraved with the number 6262, though the model is a reference 6263. There is yet more proof of the boom in the price of Rolex Daytona watches: the almost irrational soaring of the price of a 1969 'Paul Newman Oyster Sotto' reference 6263 model, featuring a sublime chocolate-coloured dial, which was sold for £1,520,500 ($1,952,300) by Aurel Bacs at Phillips in May 2016.

PATEK PHILIPPE REFERENCE 1518 IN STEEL, 1943

A SAGA
NAMED '1518'

This extremely rare Patek Philippe reference 1518 in steel is a real icon. It was sold at auction by Phillips in Geneva in November 2016 for nearly £9 million ($11.5 million), beating all records for its category. As of today, it is the most expensive Patek Philippe wristwatch in the world – quite simply because it embodies, better than any other, the quintessence of Patek Philippe.

Patek Philippe,
reference 1518 dating
from 1943: steel case,
35mm in diameter,
hand-wound
mechanical movement,
chronograph and
perpetual calendar
(130 Q), Gay Frères
'beads of rice' bracelet
and folding clasp.

The number 1518 is a secret code, a divine figure, almost a magic formula. For several decades, these four digits have made collectors dream. They are the secret fantasy of auctioneers in the most prestigious auction houses. Produced for only 14 years, between 1941 and the early 1950s, the 281 Patek Philippe reference 1518 watches are mostly made of yellow gold. Some – about 20 per cent of the total – are made of pink gold. Even rarer are the four made out of steel – an iconoclastic metal for such a jewel. After all, the reference 1518 is no ordinary watch. It combines, for the first time in history, two complications that at first glance appear incompatible: a sports-inspired chronograph and a perpetual calendar, synonymous with technical greatness. At the time, Patek Philippe's watchmakers were the only ones capable of bringing these two major functions together in a single wristwatch case. Specialists such as John Goldberger, author of an authoritative book on Patek Philippe steel watches, argue that the Geneva manufacturer was literally financially reborn because of this reference 1518. As a result, when one of these models is put up for auction, the collecting community holds its breath – especially when the watch in question is a 1943 steel version. This was what happened at a Phillips auction in Geneva in late November 2016. After more than 13 minutes of madness, in front of 400 people in the room, with 500 more on the telephone, the expert Aurel Bacs sold this little marvel for nearly £9 million ($11.5 million). The world record for a wristwatch had been broken. The original owner, the Hungarian Joseph Lang, had bought the watch in Budapest on 22 February 1944. He paid 2,265 Swiss francs. In fact, he bought two of these watches that day – exactly the same reference, both in steel. It is still not known why he purchased two but only kept one. These two watches resurfaced in Hungary in the late 1990s. One of them was sold. The other is probably stowed away in the safe of a knowledgeable enthusiast or a ferocious speculator.

PATEK PHILIPPE REFERENCE 1527,

PERPETUAL CALENDAR AND CHRONOGRAPH, 1943

■

THE PATEK PHILIPPE THAT GOES AGAINST THE FLOW

A mysterious history and a unique personality: this chronograph, reference 1527, manufactured in 1943, with a perpetual calendar and moon phase indicator, is one of the most incredible Patek Philippe watches ever made. Probably the only example produced, and completely against the aesthetic norms of the 1940s, the reference 1527 is, as of today, one of the most expensive Patek Philippe wristwatches in history.

In 1941, at the Basel Watch Fair, Patek Philippe presented to the public for the first time one of the legendary models that established its reputation. The reference 1518 was the first production-line wristwatch featuring a perpetual calendar (date, day, month and moon phase, automatically allowing for the number of days in the month and for leap years) and a chronograph. Beyond the technological innovation involved in the total integration of these complications within a hand-wound mechanical movement, Patek Philippe used this model to define a conventional and characteristic style. The watch's golden case, with a diameter of about 35mm, symbolizes watchmaking classicism par excellence, the perpetual calendar being a high-quality complication of Patek Philippe's.

In 1951 the reference 1518 was replaced by the more modern reference 2499: its design showed radical modifications that are still found on some of the brand's models today.

In 2004 the Patek Philippe Museum caused great excitement by displaying a reference 1527 dating from the early 1940s, which had belonged to Henri Stern, grandfather of Thierry Stern, the company's current president. It was a gold watch with a diameter of 37.6mm, featuring a perpetual calendar and moon phase indicator. Its design, and the large size of its case – almost out of proportion for the time – are groundbreaking compared to those of the traditional reference 1518 watches. A few years later, watchmaking historians discovered that it had been a special order by Charles Stern, one of the major figures in the family's history. When he died in 1944, it was inherited by Henri Stern; and when he in turn died in 2002, it reverted by right to Philippe Stern, who wore it for two years before entrusting it to the company's museum.

For a long time it was believed that this reference 1527 was unique. However, a second watch bearing the same reference number does exist; it was bought for £4,794,400 ($6,156,000) on 10 May 2010 at a memorable Christie's auction in Geneva. As well as the perpetual calendar and moon phase indicator, this reference 1527, which was acquired by a private collector, features a chronograph. This 'QP' (which stands for quantième perpetuel – French for 'perpetual calendar') in pink gold is, as of today, the third most expensive Patek Philippe watch in the world.

PANERAI RADIOMIR 'EGIZIANO', 1956

ON EGYPTIAN TIME

In 1956, at the request of the Egyptian navy, the Florentine company Panerai produced a limited edition of a diving watch. This model was as rare as it was legendary, notably on account of its size (its diameter was 60mm), which was regarded as excessive at the time. The 'Radiomir Egiziano' is the holy grail of the world's *paneristi*.

It all began in the mid-1950s, when the military situation in the Middle East was extremely tense. Egyptian Naval Commander Fawzi asked Panerai, a Florentine company that had specialized in the design and manufacture of professional diving instruments and watches since 1936, to produce some 50 diving watches. They were finished in February 1956 and bore the reference GPF 2/56, the company's internal code that included the month and year of launch. The watch was named Radiomir Egiziano – the 'Egyptian Radiomir'.

In December 2014, at the end of the Artcurial's 'Panerai Only' monothematic auction, an extremely rare Radiomir Egiziano GPF 2/56 bearing the serial number N.E. 007 was snapped up for £110,500 ($141,900), a world record for this model; ever since, it has been stored in the safe of a major Asian collector. The auction catalogue explains: 'It is exceptional in design and, with its 60mm diameter, was at the time the biggest wristwatch in the history of watchmaking.' But it is also the first and only professional diving watch with a movement that has an eight-day power reserve. Its bezel is unique and groundbreaking: it is bidirectional and features a system of micro-ball bearings that click as it rotates. Even the superstar of diving watches, the 'Submariner', would be fitted with this kind of bezel only in the 1980s, some 30 years later. This makes the GPF 2/56 a real museum piece, all the more so as it bears one of the nicest numbers – 007. With all its avant-garde technology, this watch would have made the perfect accessory for the famous British spy.

The interest Panerai collectors show in the Egiziano is entirely justified: this watch brings together all the brand's DNA. It has a large circular case with a back secured by six screws and, above all, it features the first crown-protecting bridge, patented in the same year, 1956, which became so characteristic of Panerai. It also has a black 'sandwich' dial – so called on account of the cut-out numerals and indexes, which enable the luminous substance beneath the dial to shine through. On the dial, at 3 o'clock, opposite the small second hand at 9 o'clock, a circle simulating a subdial is surrounded by the inscription '*8 Giorni Brevettato*' ('patented eight days'), which gives the length of the movement's power reserve. Its dome-shaped glass – in Plexiglas – is ingeniously protected by a bidirectional revolving bezel that is graduated with screws. When it is operated, the

THE BIGGEST WRISTWATCH EVER MADE

screws click using a mechanism of three micro-ball bearings located in the case middle at 1 o'clock, 5 o'clock and 9 o'clock. At the time, this innovation was unique. 'It is not known whether 50 really exist; it is unlikely,' notes Artcurial's watchmaking department. 'What is certain is that, over the past 15 years, fewer than 10 of these watches, including this one, have come up for sale at auction. These bore the numbers N.E. 012, N.E. 038, N.E. 018, N.E. 032, N.E. 002 and N.E. 040. The oldest ones were fitted with the Radiomir and the more recent ones with the Luminor dial.' Panerai has produced re-editions of the Egiziano, but none of these modern watches can possess the charm of the very first ones ever made.

PORTFOLIO

■

THE GEMS OF ONLY WATCH

Since 2005, dozens of unique watches have been conceived, created and sold, thanks to Only Watch, a charitable auction organization that raises money for the fight against Duchenne muscular dystrophy, a genetic condition that causes progressive degeneration of the muscles. Organized in connection with the Association Monégasque contre les Myopathies (the Monaco association for the fight against myopathy) and the Monaco Yacht Show, under the patronage of Prince Albert of Monaco, Only Watch auctions have raised tens of millions of euros and enabled medical research, as well as ingenuity in watchmaking, to make significant progress.

MB&F HOROLOGICAL MACHINE NO. 4 THUNDERBOLT 'FLYING PANDA'

A remarkable fusion of a childhood dream and avant-garde, high-level watchmaking: that sums up the Horological Machine (or HM) No. 4 Thunderbolt 'Flying Panda', conceived by Maximilian Büsser for the 2011 Only Watch auction. A few years earlier, Maximilian Büsser, founder of MB&F, had been struck dumb in front of a painting by the Chinese artist Huang Hankang, depicting a panda on a rocket, and had bought it. The idea of associating this childlike image with his HM No. 4 Thunderbolt seemed to him a natural one. The result speaks for itself: a UFO of a watch, with 311 parts in a titanium case inspired by aviation, with its two aerodynamic nacelles and a watch movement of extreme complexity (hand-wound, with two barrels in parallel), visible through two transparent sapphire panels, which required more than 100 hours of machining. A watchmaking machine of three-dimensional kinetic dreams. Naturally, it was Huang Hankang who designed the little panda.

Only Watch 2011, £149,300 ($191,700).

CHRISTOPHE CLARET X-TREM-1 PINBALL

Christophe Claret is a unique watchmaker, always in search of innovation. Each of his pieces is a challenge to time. Designed in 2013, this unique X-TREM-1 Pinball watch draws inspiration from pinball machines: the small balls that mark the hours and minutes, which are set within sapphire tubes in the traditional versions of the X-TREM-1, have been moved here into cylindrical grilles. The flying tourbillon mechanical movement (inclined at a 30-degree angle) has also been redesigned, to make it resemble the inside of a pinball machine: bumpers, slingshot and balls are integrated within its complex mechanism. Last but not least, the inscription 'TILT' at the top allows the time to be rapidly adjusted.

Although it secured a bid of £88,000 ($113,000) at the 2013 auction, the deal was never honoured. Eventually, the watch was sold to a collector for £228,300 ($293,100).

LOUIS VUITTON ESCALE WORLDTIME 'THE WORLD IS A DANCEFLOOR'

Louis Vuitton has always offered its clients the option of personalizing their luggage with bands of colour, initials or geometric, hand-painted pictograms. It is that world of colour that inspired the creation of the Escale Worldtime watch; its dial, hand-decorated with the technique used for painting miniatures, requires 40 hours of work and the application of 30 different colours, one by one. Here, Louis Vuitton has reinvented the 'universal time' function, which makes it possible to tell the time instantly in all 24 time zones. This brightly coloured version was designed for Only Watch: a one-off piece that gets itself noticed.

Only Watch 2015, £99,600 ($127,900).

BELL & ROSS BR 01 'SKULL BRONZE TOURBILLON'

By bringing together the symbolism of the skull, the sophistication of an exceptional mechanical tourbillon movement and the amazing patina of a bronze case that looks as if it has emerged from the abyss, the French watchmaker has created here a truly emblematic piece. The BR 01 'Skull Bronze Tourbillon' watch has a square 46mm case in antique-finish bronze, a hand-carved skull in solid gold, hand-wound movement with four complications (tourbillon, regulator, precision indicator and power reserve indicator) and an patinated alligator strap. It is inspired by the legend of the shipwreck of the *Black Wind*, the ship of the pirate Bartholomew Hawkins, which sank off Monaco in 1815.

Only Watch 2015, £76,600 ($98,400).

F. P. JOURNE 'TOURBILLON SOUVERAIN BLEU'

Taking part for the first time in an Only Watch auction in 2015, François-Paul Journe pulled out all the stops and created a unique piece, with an exceptional design and mechanism. This was the first time a tourbillon was set within an F. P. Journe case made entirely of tantalum, a metal that is notoriously hard to work and owes its name to the famous 'torment of Tantalus' of Greek mythology. The mesmerizing chrome blue dial reflects light like a mirror. It is a unique watch in tantalum, with an F. P. Journe movement in pink gold, featuring a remontoir (constant force device) and the dead-beat seconds complication.

Only Watch 2015, £421,300 ($541,000).

VAN CLEEF & ARPELS POETIC COMPLICATION WATCH 'DE LA TERRE À LA LUNE'

The undisputed master of 'poetic complications', Van Cleef & Arpels conceives and designs rare watches that invite you on imaginary journeys. For the 2011 Only Watch auction, the jewellery manufacturer took inspiration from the 19th-century French writer Jules Verne's *Voyages Extraordinaires* to create a unique piece with great power to evoke dreams. The principle of the retrograde movement, which allows the two hands to describe circular arcs of 120 degrees on the dial and return immediately to their starting position, is here wrapped in poetry. On the right, the rocket Jules Verne imagined is heading for the moon, and marks the minutes from 0 to 60. On the left, a star marks the hours from 0 to 12. On the dial, the sky is dotted with stars and planets inlaid with semiprecious stones and *champlevé* enamel.

Only Watch 2011, £188,800 ($242,400).

PIAGET ALTIPLANO 900P

Piaget's Altiplano watch embodies the manufacturer's motto in its genes and in its shape: 'Always do better than necessary.' A conventional watch generally contains, in its case, a movement 'fixed' onto a plate. Here, the hand-wound calibre is merged with the bottom of the case. It is neither entirely a movement, nor entirely a case – instead, it is both at once. The Altiplano 38mm 900P is the world's thinnest mechanical watch, at just 3.65mm. For Only Watch's 2015 auction, Piaget produced this variant – a unique version with red hands. 'Each of the 145 parts composing the Altiplano 38mm 900P has been trimmed to a size sometimes barely thicker than a hair's breadth – including some wheels measuring a mere 0.12mm thin,' Piaget explains.

Only Watch 2015, £53,600 ($68,800).

PATEK PHILIPPE, REFERENCE 5016A-010

The Patek Philippe reference 5016 encloses in its very classic case the three complications that are most sought-after by collectors: a tourbillon, minute-repeater and perpetual calendar, with moon phase indicator. This watch contains 506 components, features an elegant enamel dial and is a key piece in the world of grand complication watches. It also features a retrograde date hand, which returns to 1 after reaching the 28th, 29th, 30th or 31st of the month via an ingenious mechanism. Created especially for Only Watch's 2015 auction, reference 5016A-010 is the first – and only – version of this watch ever produced in steel. As a result, it broke all records, finding a buyer at £5.6 million ($7.2 million), the highest price achieved by a wristwatch since the charity auction's launch in 2005. As of today, it is the second most expensive Patek Philippe wristwatch in the world.

Only Watch 2015, £5,591,800 ($7,179,900).

PATEK PHILIPPE, REFERENCE 5004T-001

At Patek Philippe, grand complications symbolize the quality of the craft and are therefore made from high-quality materials: platinum, or yellow, grey or pink gold. With some exceptions, using steel is out of the question. As for titanium, don't even think about it – except for the Only Watch auction. Patek Philippe's reference 5004 is the ultimate hand-wound classic chronograph. Its calibre of 407 parts is based on a Nouvelle Lémania ébauche (blank movement) that was completely reworked in the manufacturer's workshops. Patek Philippe's watchmakers added a flyback hand and perpetual calendar mechanism. They daringly made the case of this reference 5004T-001 watch, specially created for the 2013 Only Watch auction, out of titanium. It is the only version ever made from this ultralight material.

Only Watch 2013, £2,590,100 ($3,325,700).

PATEK PHILIPPE 'CALIBRE 89', 1989

■

150 YEARS OF EXPERTISE

In 1989, for its 150th anniversary, Patek Philippe made a pocket watch with 33 complications bearing the Poinçon de Genève, a certificate awarded to timepieces of remarkable quality and craftsmanship. Until the appearance of the Vacheron Constantin reference 57260 in 2015, the 'Calibre 89' retained the coveted title of 'the world's most complicated watch'. In Geneva in 2009 the yellow gold version, one of only four existing pieces, sold for more than £3.5 million ($4.5 million) at an unforgettable Antiquorum auction.

For this renowned Geneva manufacturer, each major anniversary is an opportunity to create an exceptional watch, generally produced as an ultra-limited edition. It is a fitting way to celebrate its rich history – and to prove, each time, its supremacy in the world of grand complications.

In 1989, its 150th anniversary, Patek Philippe caused a minor earthquake by bringing out its 'Calibre 89', a pocket watch of extreme complexity, featuring 33 complications, which was a distillation of all the company's expertise. Only four were made (in platinum and in pink, yellow and white gold). This watch is a champion performer in every department: it has 1,728 components and 33 complications contained in a case 88.2mm in diameter and 41.07mm thick, weighing 1.1kg (2.4lb) in total. The research and development for this masterpiece took five years and its manufacture took another four: nine years of full-time work for nine engineers and watchmakers. And all without any help from computers: preliminary plans, sketches, detailed design and drawings were all done on a drawing board.

The 33 complications are contained in two dials that feature eight discs and 24 hands. The 'Calibre 89' is probably the watch that comes closest to perfection, displaying sunrise and sunset time in Geneva, month, season, decade, century, age and phases of the moon, *grande sonnerie, petite sonnerie*, chronograph, flyback hand, Gregorian calendar, star chart at the latitude of Geneva – and the date of next Easter, which appears on 31 December at midnight.

The last of these functions, which was patented in 1985 (patent no. 649673) is a real rarity: the 'Calibre 89' is the only watch in the world that features it. Easter is a moveable feast: in the Gregorian calendar, Easter Sunday can fall anywhere between 22 March and 25 April. The mechanism invented by Patek Philippe is singular, according to the patent, in that it drives 'a succession of operations while keeping an action on hold for the time required for the others to take place. This watchmaking achievement, which is remarkable in an age dominated by electronics, reflects all the potential of mechanical watches.' Rumour has it that the four watches were bought in 1989 by the same royal family. The complication indicating the Easter date was, apparently, a decisive factor at the time for the purchases. Today, the collection has been dispersed.

ROLEX COSMOGRAPH OYSTER DAYTONA

PAUL NEWMAN 'RCO', 1969

■

THE CHRONOGRAPH
EVERYONE DREAMS OF

Few watches create as much passion as the hand-wound Rolex Daytona vintage chronograph.
A safe investment for collectors and the star choice for counterfeiters, this model, with its sporting
style, has spawned an abundance of literature. The rare 'Paul Newman' version, with its distinctive
dial featuring two or three colours, is the most sought-after. In Geneva in November 2013 a model
featuring an anomaly on its dial known as 'RCO' was snapped up for more than $1 million (£780,000)
at a Christie's auction.

Created in 1963, this hand-wound mechanical chronograph, with a design inspired
by the world of motor racing, cost barely $210 plus tax when it was introduced
onto the American market. Those models were made in small numbers because, at the
time, they did not meet with much success.

Several references (6238, 6239, 6240, 6241, 6262, 6263, 6264 and 6265) were produced
between 1963 and 1987. Each had its own specific technical and aesthetic details:
the typeface of the indexes, the word 'oyster' sometimes present on the dial and
sometimes not, 'millerighe' screw-down push buttons with fluting of different widths,
different dial colours and so on.

Certain references are even higher quality than others, such as the 'Paul Newman'
models – the unofficial name of references 6239, 6241, 6262, 6263, 6264 and 6265 –
which have subdials in a different colour from the main dial. These 'Paul Newman'
references feature a dial with two or three colours (also known as an exotic dial). Rolex
never made the name official, but numerous photographs demonstrate how the
American actor, a fan of motor sport, wore this watch almost every day, fitted with
a leather strap. It had been given to him by his wife, Joanne Woodward; on the back
of the casing she had the following words engraved: 'DRIVE SLOWLY ME'.

Unlike generic Daytona watches, a 'Paul Newman' uses an Art Deco typeface for the
figures on its dials. The outside indexes are highlighted with a line in a contrasting colour
to the dial. There is also a small 'sight' in each subdial, of which certain indexes end
with a square. Finally, the hour indexes on the subdial at 9 o'clock show the figures
15, 30, 45 and 60, whereas on regular Daytona watches there are only the figures 20,
40 and 60. When a piece with a 'quirk' comes on the market, things start to go crazy.
This was the case at the Christie's auction held in Geneva on 10 May 2013: a 1969
Daytona Paul Newman with a black dial, known as 'RCO', sold for £757,600 ($972,800)
in a supercharged atmosphere.

Most Daytona watches bear the inscription 'Rolex Oyster Cosmograph' – that is, 'ROC';
the 'RCO' model bears the inscription 'Rolex Cosmograph Oyster'. The transposition
of the words 'Cosmograph' and 'Oyster' changes everything. Only 12 pieces in the
world feature this oddity, which is very probably the result of an error during
the manufacture of the dial.

PATEK PHILIPPE DOCTOR'S WATCH, REFERENCE 130, IN STEEL, 1937

■

A DELUXE MEDICAL INSTRUMENT

Only two of these Patek Philippe single-button chronographs in steel with a pulsometer scale exist. The first is in the collections of the Patek Philippe Museum in Geneva; the second found a buyer on 10 May 2015 for the improbable sum of 4,645,000 Swiss Francs (£3,256,000/$5,019,000), at the conclusion of an unforgettable Phillips auction. It is extremely rare and one of the most expensive Patek Philippe steel wristwatches in the world.

Wednesday 24 March 1937 was an important day for Walser, Wald & Cia, Patek Philippe's distributor in Buenos Aires. The company took delivery of two exceptional watches, probably ordered by two wealthy brothers, both doctors. These pieces, reference 130, barely resemble other watches of the day. They have a 35mm-diameter Staybrite steel casing – at a time when the custom was to wear smaller, gold models. These two watches, numbered 504'146 and 504'147, have a medical vocation – they are, in short, professional watches. They feature a single-button chronograph with a movement designed by the watchmaker Victorin Piguet in the small Swiss village of Le Sentier, and are among the most complicated watches made by the Geneva watchmaking industry at the time. Their matt silver dial features a pulsometer scale calibrated for 30 pulsations, which allows a patient's heart rate to be rapidly calculated, as well as the chronograph's minute totalizer, located at 12 o'clock. Patek Philippe had already made several doctor's watches with the same reference and design, but almost all were in gold with an 'officer' type casing, 33mm in diameter – a size more in keeping with the fashion of the time. This meant that the appearance of the reference 130, with its very rare steel casing and 'oversize' diameter, at a Phillips auction in May 2015, was a big event. The highest estimates were as much as £1.5 million ($1.9 million); but the atmosphere at Phillips that day was electric and Aurel Bacs's hammer reached unexpected heights. This reference 130 – in exceptional condition, never restored, with its original buckle and only two owners – smashed all records, doubling its most optimistic estimate and selling for more than 4.5 million Swiss Francs.

■

SWISS MOTHER, FRENCH FATHE PORTUGUESE BY ADOPTION

Now in the Musée du Temps (Museum of Time) in the heart of the Palais Granvelle, Besançon, France, the Leroy 01 was the most complicated watch in the world for almost a century. Its movement comprises 975 components on four different levels and took seven years to perfect. It features 26 indications – most directly concerning the measurement of time – and a casing in 18-carat yellow gold, 71mm in diameter, with an overall weight of 228g (8oz). It is one of the most impressive watches ever made.

In Lisbon at the end of the 19th century, the Portuguese António Augusto de Carvalho Monteiro, a great enthusiast of high-level watchmaking, placed a private order with the house of Leroy. A famous entomologist, he had made his fortune in coffee and precious stones, and was an enlightened philanthropist. He wanted to treat himself to a unique piece that brought together everything science and mechanics could fit into a watch. Leroy, founded in 1751 by Basile Le Roy, was by then one of France's best-known watchmakers. Work began on Carvalho's order on 1 November 1897. The movement, comprising 975 components, was devised and fine-tuned by Charles Piguet in the small, remote village of Le Brassus, Switzerland. It took seven years of research and work to achieve his goal. The watch, named Leroy 01, was then assembled and adjusted in the workshop of Louis Leroy in Saint-Amour square, Besançon.

With between 20 and 26 complications (experts disagree on the number) contained within a finely engraved and carved 71mm-diameter yellow gold casing, the Leroy 01 has a dial on each side. It displays the time in 125 cities, the seasons, solstices and equinoxes, and the sky above Lisbon, Paris and Rio de Janeiro. It features a perpetual calendar, hygrometer, barometer, altimeter and compass – cleverly set within a hollow in the crown.

In 1900 the Leroy 01 was still at the development stage, but that didn't stop Louis Leroy exhibiting this jewel of technical prowess at the Exposition Universelle in Paris, where it was awarded the top prize for watchmaking. The watch was delivered to its owner on 15 November 1904. 'Out of respect for all the protagonists in this unique adventure, which has been crowned with success after seven years of work, we like to say that this piece was born of a French father and a Swiss mother,' Leroy explains today, not without a touch of pride.

The Leroy 01 remained in the Portuguese collector's family until 1953. Thanks to a subscription committee who raised more than 2 million old French francs, this 228g (8oz) treasure ended up in the collections of the Musée du Temps. Today, it is displayed with its original ebony casket containing spare dials, a set of hands and various other components. The Leroy 01 remained a record-holder in watchmaking technical prowess until 1933, when Patek Philippe's 'Henry Graves Jr Supercomplication' was created.

PATEK PHILIPPE, REFERENCE 2458

OBSERVATORY CHRONOMETER FOR J B CHAMPION, 1952

■

A PATEK PHILIPPE
FOR A CHAMPION

In 2012 this 'relatively simple' Patek Philippe broke all records when it sold for almost £3 million ($3.8 million) at a Christie's auction in Geneva in 2012. Coming from a well-known American collection, it fascinated experts and collectors, as do all pieces bearing the words 'Made especially for J. B. Champion'.

J B Champion Jr, also known as Joe Ben, was one of the greatest American lawyers of the 20th century. He was also one of the greatest watch collectors of his generation. His personal collection – comprising pieces only by Patek Philippe and Vacheron Constantin – has never been precisely catalogued, but his passion for rare Patek Philippe watches was legendary. His love of fine watchmaking sometimes verged on the eccentric: every morning he took his watches from the safe and wound them up one by one. If, at the end of the day, one of them showed a chronometric deviation, he sent it back at once to Linz Bros or the Henry Stern Watch Agency, his favourite New York retailers. He was even known to examine the movements of his rarest models under the microscope. J B Champion Jr loved special orders, only in yellow gold, pink gold or platinum and, with slightly egocentric vanity, had the words 'Made especially for J. B. Champion' inscribed on their dials.

This watch, with its a 36mm platinum casing, a movement that was twice awarded the Poinçon de Genève and a calibre that came third in the Geneva observatory chronometer competition, was put up for sale at Christie's in Geneva on 12 November 2012, accompanied by an extra dial and a platinum bracelet. This auction also featured an extremely rare Patek Philippe chronograph, reference 2499, in platinum with a perpetual calendar, put up for sale by Eric Clapton. The 'J B Champion' wasn't expected to stand much of a chance against the English singer's little jewel, and yet it broke all records and sold for £2,894,700 ($2,716,800). The buyer was Alfredo Paramico, a famous collector who has since sold it on. It was reportedly bought by Angelina Jolie, who supposedly gave it to Brad Pitt – a rumour that has only further increased the appeal of this unique piece.

WATCHMAKING
AT THE
HIGHEST
LEVEL

FRANCK MULLER AETERNITAS MEGA 4

■

ONE WATCH,
36 COMPLICATIONS

In 2009 Franck Muller revealed the Aeternitas Mega 4 – an exceptional piece comprising
1,483 components. It took five years of research and development and features 36 complications,
a world record for a wristwatch.

The project was an insane one: to fit 1,483 components on a movement plate 34.4mm by 41.4mm, itself contained in a Curvex-shaped grey gold casing, 42mm wide, 61mm long and 19.15mm thick. Trickier still, the aim was to fit 36 complications in this minuscule space. A mad idea – which is why Franck Muller had set his heart on accomplishing it.

The Aeternitas Mega 4 wristwatch brings together everything that makes real watchmaking enthusiasts dream: a *grande sonnerie*, a minute-repeater, a single-button chronograph with flyback hand, a secular perpetual calendar, an equation of time, a *petite sonnerie* with Westminster chimes, a tourbillon and astronomical moon phases that are only 6.8 seconds out every lunar month – that is, about a day over 1,000 years. The very first example was bought for $2.7 million (£2.1 million) by an American collector, Michael J Gould. The watch was presented to him in the Empire Salon of the Hôtel de Paris, Monaco.

THE 36 COMPLICATIONS OF
THE AETERNITAS MEGA 4

1   Day/night indicator

2   *Grande sonnerie*

3   *Petite sonnerie*

4   Silent mode

5   Minute-repeater

6   Westminster chime with four gongs

7   Programming of the *grande sonnerie* or *petite sonnerie*
    function via a water-resistant button with display on the dial

8   Programming of the *sonnerie* or silent mode function via
    a water-resistant button with display on the dial

9   Mechanism for isolating the *sonnerie* when the time is
    being set

10 Mechanism preventing the sounding of a new *sonnerie* if the
    preceding one has not finished

11 Bolting mechanism for setting the time while the *sonnerie*
    is sounding

12 Transmission mechanism to the hammers, allowing
    adaptation to different shape of the gong

13 Movement power reserve indicator (three days)

14 *Sonnerie* power reserve indicator (36 hours)

15 Silent centrifugal regulator of the *sonnerie* rate of striking

16 Flying tourbillon on a bearing with ceramic balls

17 Balance with gold adjustment screws, without index

18 Breguet overcoil with Phillips curve

19 Flying tourbillon cage, without bridge, visible in the dial

20 Automatic self-winding of the movement via a platinum
    micro-rotor

21 Automatic self-winding of the Westminster chime via
    a platinum micro-rotor

22 Perpetual calendar

23 Day display

24 Month display

25 Retrograde date display

26 Secular calendar

27 Year display up to 999 years

28 Leap year display

29 Secular year display

30 Astronomical moon phases: only 6.8 seconds out every lunar
    month – that is, one day over 1,000 years

31 Equation of time

32 Two additional, independent time zones

33 Integrated chronograph with three column wheels

34 Instantaneous minute counter

35 Hour counter integrated with the chronograph mechanism,
    retrograde display

36 Flyback hand mechanism

JAEGER-LECOULTRE HYBRIS MECHANICA À GRANDE SONNERIE

■

JAEGER-LECOULTRE ON ENGLISH TIME

In 2009 Jaeger-LeCoultre released a series of three exceptional watches that brought together all the company's expertise: the Hybris Mechanica à Grande Sonnerie, the Hybris Mechanica à Triptyque and the Hybris Mechanica à Gyrotourbillon. These timepieces, of which 30 were made and sold in Alcantara leather caskets, are a remarkable watchmaking trilogy totalling no fewer than 55 complications. With its 1,300 components and miniature Westminster chime, the Hybris Mechanica à Grande Sonnerie is the most impressive of the three.

The almost magical mechanism in the Hybris Mechanica à Grande Sonnerie is probably one of the most highly sophisticated ever made for a wristwatch. Here are a few numbers: the Jaeger-LeCoultre 182 calibre is 10.42mm thick and 37mm in diameter; it comprises 1,300 components and 26 complications, and is contained in a grey gold casing 44mm in diameter and just 15mm thick.

As well as a perpetual calendar with retrograde hands, flying tourbillon and jumping hour movement, the Hybris Mechanica possesses an unparalleled musical element, with three sonnerie modes: *grande sonnerie*, *petite sonnerie* and silent.

It is the *grande sonnerie* with Westminster chime that is the most astonishing. As well as sounding the hours, quarter-hours and minutes at the press of a button, like a minute-repeater, the *grande sonnerie* also automatically sounds the passing of hours and quarter hours. This extraordinarily complex watch is the only one that faithfully reproduces the tune of Big Ben. It also plays the longest melody ever heard on a wristwatch. To perfect this masterpiece, the watchmakers of Le Sentier, the little Swiss village where the Jaeger-LeCoultre workshop is based, had to redouble their ingenuity, notably redesigning the hammers and gongs, which are made out of special alloys. The price of this luxury musical instrument? £1.8 million ($2.3 million).

Seen from the back, the Hybris Mechanica à Grande Sonnerie is impressive. Its mechanism, 10.42mm thick and 37mm in diameter, contains some 1,300 components.

PATEK PHILIPPE GRANDMASTER CHIME, REFERENCE 5175R

THE GREAT ART OF GRAND COMPLICATIONS

At Patek Philippe, every anniversary is an opportunity to create an exceptional watch. For its 175th, in 2014, the famous Geneva watchmakers brought out the Grandmaster Chime, its most complicated wristwatch, with a two-dial design, 20 complications and new-baroque look. It is a masterpiece of minute mechanical work, combining a movement of extreme complexity with a casing of great style. Only seven of this commemorative model were made.

The development, manufacture and assembly of Patek Philippe's Grandmaster Chime required more than 100,000 hours' work, of which 60 per cent was spent just designing its movement, which has 1,366 components. The casing alone has 214. This meant that the seven examples made of this commemorative piece required the manufacture of 11,060 components, which were painstakingly assembled by hand. That is the price of excellence. Beyond the numbers, the Patek Philippe Grandmaster Chime is a declaration, the assertion of supremacy in the field of (very) high-level watchmaking and grand complications.

Its two-faced casing houses four barrels and 20 further complications, including a *grande sonnerie* and *petite sonnerie*, minute-repeater, instantaneous perpetual calendar displaying the year with four digits, a second time zone and an acoustic alarm that chimes the time when it goes off, as well as a date repeater that sounds the calendar if desired – two patented world firsts. Equally impressive is the double-faced 47mm casing, made possible by an ingenious rotation mechanism housed in the lugs that attach it to the bracelet.

The beauty and significance of this £1.9 million ($2,439,600) watch lie chiefly in what it symbolizes. Twenty complications means four more than on the 2005 Vacheron Constantin 'Tour de l'Île', brought out for the company's 250th anniversary, but 16 fewer than on Franck Muller's Aeternitas Mega 4; for Patek Philippe, technical considerations are not the absolute criterion of quality. This very old watchmaking company also likes to emphasize its independence and the fact that it is a family business. And in that respect, Patek Philippe comes out on top: Vacheron Constantin and Jaeger-LeCoultre belong to the Richemont international group and Franck Muller has been an independent and singular watch brand only since 1991.

TWO FACES, 20 COMPLICATIONS, 100,000 HOURS' WORK

The two faces of the Grandmaster Chime reference 5175R's pink gold casing, decorated with a hand-engraved leaf motif, are devoted to two functions: the time and the calendar. However, these two faces are the same shape, enabling the watch to be worn with either dial displayed. The reference 5175R is also a peak of ingenuity: with a chiming mechanism isolator display it indicates when it is safe to wind the movement to avoid damaging the intricate mechanism. Six of these watches have been sold; the seventh is in the collections of the Patek Philippe Museum in Geneva.

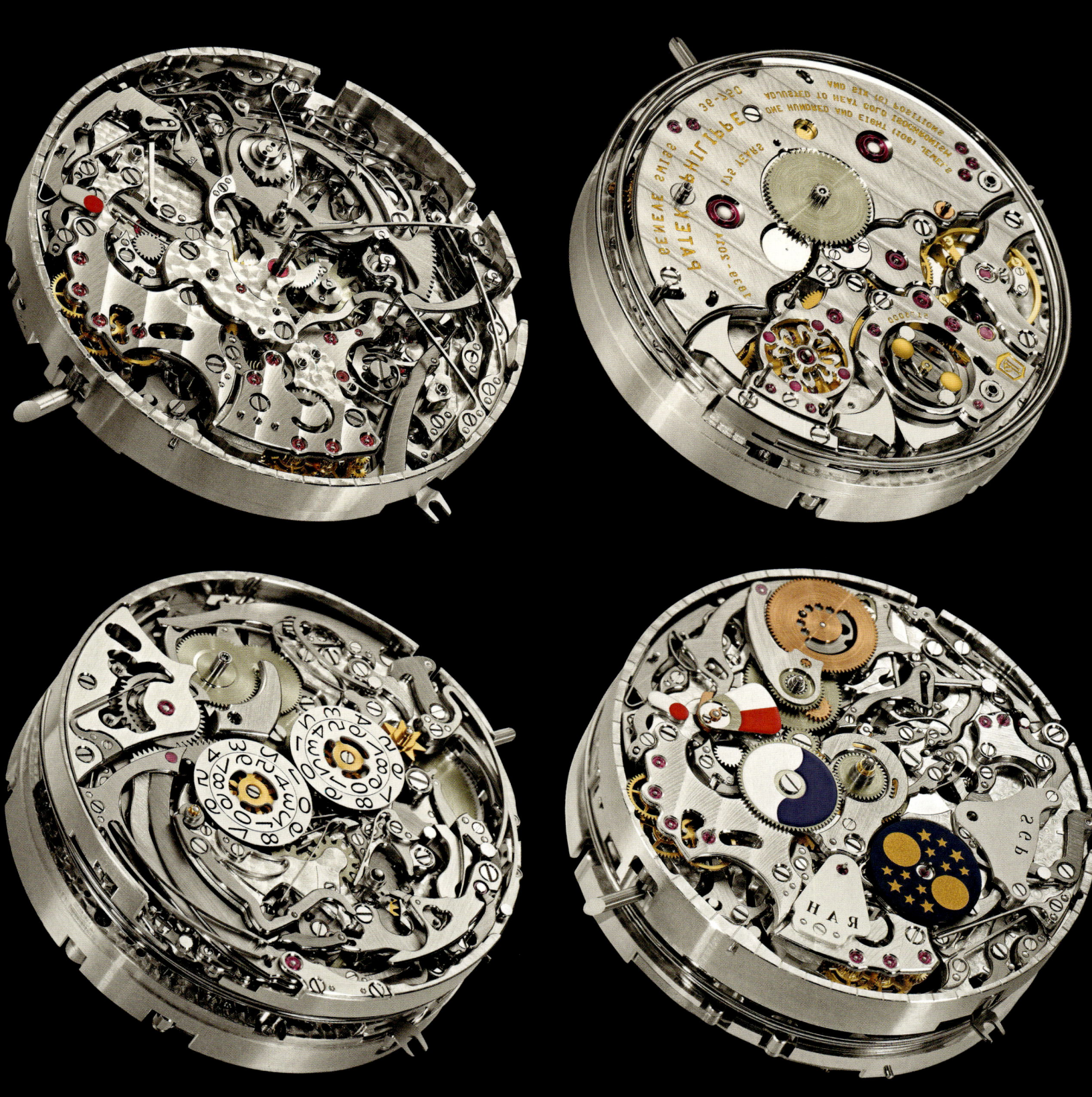

VACHERON CONSTANTIN REFERENCE 57260

■

THE WATCH THAT BROKE ALL THE RECORDS

Its elements are a concentration of all human genius and vanity. With 57 complications, this Vacheron Constantin watch is, quite simply, the most complex ever made by human hands. A one-off order by a great collector, born of a dream, it unquestionably marks a major milestone in the history of high-level watchmaking.

Vacheron Constantin celebrated its 260th anniversary in style by delivering an order a client had placed almost ten years earlier. That was the time required for three master watchmakers to come up with this dream of a watch – the most complicated ever made, with no fewer than 57 complications. These include a Hebrew perpetual calendar, a flyback chronograph with double retrograde display, a spherical tourbillon and 53 other complications that had been ordered by the watch's future owner.

The creation of custom complicated watches is a tradition at Vacheron Constantin, and reference 57260 is part of a line of exceptional creations that have punctuated the company's history. The aesthetic excellence of the manufacture is shown in the respect for, and mastery of, the demanding standards of the Poinçon de Genève. With its two dials, the reference 57260 has an elegant display. Due to the large number of complications featured, it was a challenge for the front and rear dials to offer exceptional readability. The white gold casing, perfectly balanced, has push buttons discreetly integrated into its bulk. The alarm's winding crown leaves the pure lines of the elegant casing intact, while being unobtrusive, thanks to its sophisticated concealment system.

With its two dials, the reference 57260 combines the principles of traditional watchmaking with cutting-edge 21ˢᵗ century technology. It offers new and unique complications, including several calendars and a flyback chronograph with double retrograde display. Its movement is totally innovative; the most classic complications have been modified, reinterpreted and redesigned so that this exceptional watch constitutes a perfectly harmonious whole.

More than 2,800 components were entirely decorated by hand by a single master watchmaker, employing traditional techniques such as chamfering, circular graining, Geneva Stripes and so on. The most spectacular displays include: six time measurement functions, fifteen perpetual calendar functions (Gregorian and Hebrew), nine astronomical calendar functions, one lunar calendar function, one religious calendar function, four chronograph functions, seven alarm functions and eight Westminster carillon striking functions.

TECHNICALLY IMPRESSIVE AND VISUALLY SPECTACULAR

The cage of the armillary sphere tourbillon is worth noting; it contains the watch's escapement and is visible through an opening below the star chart. Developed and manufactured by Vacheron Constantin, the reference 57260 has a power reserve of around 60 hours. It is 72mm in diameter and 36mm thick, with a total weight of 960g (33¾oz).

HARRY WINSTON HISTOIRE DE TOURBILLON 6

HARRY WINSTON'S BIG STORY

Launched in 2009 by Harry Winston, the Histoire de Tourbillon collection of limited editions is the only one of its kind. The idea was to reinterpret the tourbillon, one of the finest complications in watchmaking, in a series of spectacular and innovative shapes. The Histoire de Tourbillon 6 model, which came out in 2015, features a mind-boggling combination of multiple and inclined axes of rotation, and is the most complex ever produced.

With Harry Winston, everything is at once very simple and very complicated. The Histoire de Tourbillon 6 has an imposing white gold casing (55mm by 49mm) with a deconstructed design: its outline is dictated by a uniquely structured movement (calibre HW4701) and combines two independent time displays – one regulated by a tri-axial tourbillon and the other by a carousel. It is unusual to adapt the shape of the casing to fit the demands of a movement; usually, the mechanism fits somehow within a round or rectangular casing. Here, the opposite is true: pure watchmaking instead takes precedence.

The tri-axial tourbillon is located at 7 o'clock and stretches the limits of possibility. The first cage contains the balance, which rotates once every 45 seconds. The latter rotates within a second cage, which in turn rotates every 75 seconds, while the whole moves within a third, spherical cage, which completes one revolution every 300 seconds, on a vertical axis. It is in this last cage that the orange second hand is located, beneath a dome of sapphire glass that offers a fascinating insight into the dance of time. Diagonally opposite, a discreet carousel revolves once every 30 seconds.

Produced as a limited edition (only 20 were made), this watch displays the hours and minutes on the left, regulated by the tri-axial tourbillon and recognizable by the orange colour of the hands against a grey background. On the right (featuring skeleton hands and a blue index on a black background) is a dial with hands that can be started, stopped and reset to zero (as in a chronograph function) via a blue ceramic push button located at 2 o'clock.

GREUBEL FORSEY QUADRUPLE TOURBILLON SECRET IN PLATINUM

■

INNOVATION IS THE MAGIC WORD

Founded in 2004 by Robert Greubel and Stephen Forsey, the Greubel Forsey company specialized from the outset in designing and manufacturing complex watches that, with boldness and technical skill, redefine the essentials of watchmaking: a perpetual calendar with equation of time, a double tourbillon at 30 degrees, a double balance at 35 degrees and, naturally, an exclusive collection of quadruple tourbillons, including a 'secret' version in platinum that symbolizes all the expertise of this young, independent maker.

Since its foundation, Greubel Forsey has made the tourbillon – one of the finest complications – its speciality. The tourbillon, which allows a watch's chronometry to be regulated, requires an excellent knowledge of the principles of traditional watchmaking. Robert Greubel and Stephen Forsey have mastered them better than anyone, and managed to use them to create pieces of unbelievable technical prowess, sometimes bordering on the limits of comprehension.

Among their many masterpieces, the Quadruple Tourbillon Secret in platinum surely holds a special place. Produced in a limited edition of just eight pieces and priced at almost £700,000 ($900,000), this watch contains two double tourbillons, each incorporating an autonomous oscillator directly linked to a spherical differential comprising 28 components. The development of this intricate project – comprising 519 components, of which 261 are accounted for by the four tourbillon cages – had one simple goal: to improve chronometric functioning.

On the front, its subtly balanced asymmetry would leave no collector worthy of the name unmoved. The main dial displays hours and minutes, while the small seconds and power reserve are lodged in a small protrusion at 2 o'clock. The other two dials (at 5 o'clock and 7 o'clock) indicate the rotation of the quadruple tourbillon. On the back, the tourbillons, which are visible under sapphire bridges, ensure the extremely high accuracy of this exceptional timepiece.

A. LANGE & SÖHNE GRAND COMPLICATION

■

THE SIMPLICITY OF HIGH-LEVEL GERMAN WATCHMAKING

Brought out in 2013 and produced in a limited edition of only six, the most complicated watch from German maker A. Lange & Söhne is called La Grande Complication – a name that marvellously sums up the exceptional character of this creation.

After Switzerland, Germany is assuredly the other great homeland of high-level watchmaking. And if only one maker had to be singled out, it would be A. Lange & Söhne. Founded in 1845 by Ferdinand Adolph Lange, expropriated after World War II, relaunched in 1990 by Walter Lange and now owned by the Richemont group, this quality watchmaker carries on Saxony's great tradition of watchmaking, with upmarket production and more than 40 calibres in its repertoire.

In 2013 A. Lange & Söhne produced a watch that made a statement, embodying the full range of its expertise and displaying the finest complications – an acoustic signal with *grande sonnerie* and *petite sonnerie*, a minute-repeater, a flyback chronograph with minute counter and split second – allowing the user to read the elapsed time to the nearest fifth of a second via the blue-tinged steel hand in the lower subdial, which rotates on its axis, jumping five times per second. There is also a perpetual calendar with a moon phase display. All this sits within a 50mm-diameter rose gold casing. The mechanical calibre of this hand-wound watch is extremely complex; it is decorated and assembled by hand, with a low-frequency escapement system that produces 18,000 alternations per hour. The dial alone – in white enamel with Arabic numerals, railway minute track and four symmetrical subdials – symbolizes the extreme simplicity of this Grand Complication. There is just one drawback: it is, perhaps, too large to be practical.

JACOB & CO. ASTRONOMIA TOURBILLON BAGUETTE

A WATCH
IN ORBIT

For more than a quarter of a century, Jacob & Co. has made extravagance the focus of its creations. Ostentation, outrageousness, energy and limitless creativity bring forth completely singular pieces like the hypnotic Astronomia Tourbillon Baguette.

It began with a dream: to represent the immensity of the universe in the tiny space of a 50mm-diameter watch casing, with a shape inspired by the Solomon R. Guggenheim Museum in New York. The next step was to transform this dream into a masterpiece of watchmaking. 'This transcendent representation of the cosmos is made up of four parts, illustrating the connection between the earth and a constantly expanding universe, as well as the relationship between time and the dance of the stars of the solar system,' explains Jacob Arabo, the head of the company. Against an astral background in aventurine, this excessive timepiece with its three-dimensional calibre has an unusual movement, placed on a series of four arms that orbit the centre of the dial. 'While the orbit makes a complete rotation around the dial every 20 minutes, the four individual satellites complete the movement with their own trajectories.' The earth is lacquered and hand-painted. The moon is a spherical diamond with 288 facets – a patented invention by Jacob & Co., dubbed the 'Jacob-Cut'. Illuminated by this diamond, the starry sky looks as though it is lit during the 60-second rotation of these two globes, which takes place around an independent axis. The whole movement sits under a hemispherical sapphire glass. Moreover, the hand-wound calibre's tourbillon feels surprisingly weightless, thanks to three axes of rotation fitted into the mechanism's architecture. It is a tour de force, all the more so since the gravitational complication, as well as making its 60-second horizontal rotation, makes a five-minute rotation around the vertical axis, and the fourth arm of the dynamic module has an ingenious display of hours and minutes.

Finally, the Astronomia Tourbillon's exclusive JCEM01 calibre is fitted with a Phillips-curve balance spring and motor barrel that pushes the limits of watchmaking ingenuity far, very far – light years – ahead. This is a limited edition of 18 pieces, worth more than £400,000 ($500,000) each. That's the allure of watchmaking.

HUBLOT ANTIKYTHERA

■

HOMAGE TO A MECHANICAL GENIUS OF ANTIQUITY

The mysterious Antikythera mechanism was discovered in Greek waters in 1901, and dates from the 2nd century BCE. As of today, it is the oldest known mechanism that used gears. It is even considered the first analogue computer, allowing the precise calculation of astronomical positions, among other things. In 2011 Hublot used it as inspiration to create a watch of extreme complexity, reinterpreting the astronomical mechanism that has long fascinated the scientific community.

A total of 82 bronze fragments, some microscopic, all eaten away by corrosion: that is all that remains today of the mechanism, which was found in 1901 in a wreck near the Greek island of Antikythera, between Kythera and Crete. This extraordinary machine was on board a 30m- (98ft-) long galley, probably shipwrecked between 87 and 60 BCE during the Hellenistic period.

Many recent scientific studies have led to the conclusion that the Antikythera mechanism was an astronomical calculator. It probably featured cogs inside a wooden box – 33 x 18cm (13 x 7in) – which was in turn held shut by two bronze plates, each engraved with inscriptions that were evidently instructions for use. The 82 pieces of this historically significant mechanical puzzle are now in the collections of the National Archaeological Museum in Athens. Scans and X-rays have revealed the presence of numerous internal cogs and gears that are invisible to the naked eye. After countless hypotheses, experts today believe that these gears were cranked by a lateral handle, although the use of a supplementary hydraulic system cannot be ruled out.

It took several decades of research to establish the purpose of this mechanism. Today, it is accepted that this incredible machine made it possible to show different solar, lunar and planetary cycles, relating them to the civil calendar of several Greek cities of the time – such as Corinth, Delphi or Olympia – in order to indicate, among other things, the dates of athletic games in these different city-states. The Antikythera mechanism may have been designed in Rhodes, where the astronomer Hipparchus of Nicaea and the philosopher Posidonius of Apameia lived, or in Syracuse, by Archimedes, the father of statics. The debate goes on.

The 'simplified and miniaturized' version of the Antikythera mechanism, produced in an ultra-limited edition of 20 pieces, paid faithful homage to the ancient mechanism. The Hublot Antikythera Sunmoon watch, with a one-minute flying tourbillon, was exhibited at Baselworld in 2013. It is a singular timepiece, comprising 295 components (compared with the 495 components and 14 functions of the Hublot calibre in the Athens museum) and featuring seven complications, including the sun and moon functions on its dial.

The fact remains that the Antikythera mechanism did not display the time – the Ancient Greeks did not have the same conception of time as we do. On the other hand, with a highly complex interplay of wheels and cogs, it could calculate an incredible number of astronomical cycles – it was even able to predict eclipses. 'The volume of astronomical data compiled to create a mathematical model able to summarize such cycles using mechanical gear trains is astonishing evidence of the conceptual abilities of the scholars and engineers of antiquity,' explains Hublot. 'Given that a computer can generate data other than that entered into it, the Antikythera "machine" really is the first mechanical computer known to man. It was a good thousand years ahead of the first astronomical clocks created on a whole different scale in the main European cities in the Middle Ages.'

REPRODUCING THE ANTIKYTHERA MASTERPIECE IN A WRISTWATCH

In homage to this slightly mad machine, the technical team at Hublot, led by Mathias Buttet, created four watches that are replicas, in miniature, of the original mechanism (comprising 495 components and 14 functions). As a nod to the present day, the movement is complemented by hours, minutes and a tourbillon escapement system, as well as a five-day power reserve. For Mathias Buttet, the real achievement was to miniaturize what the mechanics of antiquity had developed, without losing anything of the mechanism's original character.

In 2013 Hublot brought out a 'simplified' version of this watch, the Antikythera Sunmoon: a limited edition of 20 pieces inspired by the Antikythera mechanism, but in a lighter version, with a one-minute flying tourbillon, displaying 'only' the moon and sun functions on its dial.

MONTBLANC TIMEWRITER METAMORPHOSIS 1 AND 2

■

HIGH-LEVEL WATCHMAKING THAT TRANSFORMS BEFORE YOUR EYES

In 2010 Montblanc caused a stir by bringing out a completely unprecedented piece at the Salon International de la Haute Horlogerie (SIHH) in Geneva. The Metamorphosis 1 could, thanks to a simple pull-out crown, change its face and dial, and therefore its functions. In 2014 the Metamorphosis 2 refined this device, which was contained in a red gold casing of highly classic design – it was perfectly circular. This was a conjuring trick raised to the level of fine art.

A watch featuring different functions shared between two dials is in itself nothing new. But when Montblanc brought out its Metamorphosis 1 – the result of many years of development by two independent designers, Johnny Girardin and Franck Orny, under the auspices of the Institut Minerva de Recherche en Haute Horlogerie – the methods of function changes were completely re-evaluated. Shifting the pull-out crown on the left-hand side of the casing (from 10 o'clock to 8 o'clock) starts a process of metamorphosis lasting 15 seconds: the classic timepiece becomes a chronograph. It changes before your very eyes, like a stage trick. In the watch's lower half, four shutters disappear as if by magic, sliding to the left and right under the dial's central base. The same process takes place with the two shutters on the hour dial of the regulator at the 12 o'clock position. When all the shutters are open, a small dial appears at 6 o'clock, by entirely mechanical means. An opening in this disc swallows up the date hand: the rotating disc that has just appeared becomes the chronograph's minute counter.

In its classic mode, the imposing white gold casing (47mm wide) features a counter of the regulator type, a retrograde minute hand at the centre that traces a circular arc between 8 o'clock and 4 o'clock, and a large second hand (also in the centre). The lower part of the dial has a circular date display at 6 o'clock. In single-button chronograph mode, this rotating disc becomes the chronograph's minute counter. The transformation mechanism uses 50 components that move in a perfectly synchronized way. A total of 28 Metamorphosis 1 watches were made, all in white gold.

MAGICAL TECHNICAL PROWESS FOR A WATCH WITH MULTIPLE FACES

In 2014 Montblanc brought out the second version of the Metamorphosis. This features 746 components and several technical and aesthetic improvements. It retains all its functions, no matter which configuration is displayed: the date and chronograph functions continue working independently of the watch's face. If the latter is kept in chronograph mode for several days, the date continues to change in the background and is displayed with complete accuracy when the watch is returned to its classic mode. Thanks to two independent modules, it takes only five seconds to switch from one dial to the other. Last but not least, the Metamorphosis 1's oval casing is replaced by a perfectly round casing in red gold. Only 18 of this second version were made – priced at £237,000 ($304,300) each.

TOURNAIRE PARIS FOREVER

■

ALL OF PARIS
IN ONE WATCH

The Eiffel Tower, the Pont des Arts, the Arènes de Lutèce, the Gardens of the Trocadéro, the Arc de Triomphe, the Fontaine des Innocents, Place Vendôme and Notre-Dame Cathedral – this Philippe Tournaire watch gives you a tour around the centre of Paris. The main components of the Paris Forever's mechanical tourbillon movement take on the shape of the finest monuments of the City of Light. It is an outlandish watch and the only one of its kind.

It was a slightly mad project, rather like the person behind it. An ingenious jeweller, an independent mind and an iconoclastic artist, the Frenchman Philippe Tournaire is one of those unclassifiable individuals whose creations make light of time. He is famous for his architectural rings and jewellery with stark, powerful designs. In 2010 he decided to design a watch in honour of Paris, and was determined to create a piece that resembled no other: an unreal timepiece, a watch that would be an event in itself.

Philippe Tournaire wished to create an exceptional watch with a mechanical tourbillon movement made up of components, whether functional or not, that were completely manufactured by smelting and engraved, re-creating the shape of Paris's most famous monuments. The idea was to make as many components as possible, in gold, directly on and within the movement. Philippe Tournaire approached the experts at Technotime, Swiss specialists in high-level watchmaking. Together they designed the project; all that remained was to realize it. The components were worked on one by one to give them the required degree of precision. For the first time, jewellery and watchmaking were combined.

The 'TT791.00 Architecture', a 60-second tourbillon, features a double barrel and has a power reserve of five days. The cage of the tourbillon, which resembles the Arènes de Lutèce, comprises 67 components and weighs only 0.51g ($^{1}/_{100}$ oz).

Paris wasn't built in a day. Neither was Philippe Tournaire's watch. It would take his teams at Montbrison and the watchmakers at Technotime – who focused on the extreme degree of precision demanded by the manufacture of the components - more than 18 months to bring this miracle of watchmaking architecture to life. The 'TT791.00 Architecture' calibre created a new topography of Paris through some 20 major monuments: the Trocadéro gardens and their fountain in blue diamond, the Louvre, the Académie Française, the Fontaine des Innocents, the Pont-Neuf – supporting the tourbillon and made of white diamonds - the Eiffel Tower, the Arc de Triomphe,

JEWELLERY AND WATCHMAKING JOIN FORCES TO CREATE THIS PARISIAN GEM

Place Vendôme, Notre-Dame Cathedral, the Arènes de Lutèce with the axis of the watch's hands at their heart, and so on. The gold casing, 48mm in diameter and 18mm thick, weighs 160g (5½oz). It is set with 174 diamonds (1.21 carats) and contains a hand-wound mechanical movement fitted with a 60-second tourbillon at 9 o'clock. The casing is transparent at the front, back and sides, thanks to several sapphire panels cut to tolerances of a micron; it sits as if suspended on four columns that reflect different styles of classical architecture: Doric, Ionic, Tuscan and Corinthian. This wildly ostentatious and exceptional piece found a buyer at £245,800 ($315,600). Apparently a second example is still awaiting a buyer.

CARTIER ID ONE AND ID TWO

■

IN SEARCH OF THE
WATCHMAKING OF THE FUTURE

ID One (2009) and ID Two (2012) are concept watches – ultra-sophisticated prototypes dreamed up by Cartier's Research and Development department. These unique pieces send out a powerful message: they represent what should be the precision watchmaking of tomorrow – and the day after tomorrow. They tell the time, to be sure, but they are also writing the future.

The watches could have gone by the name 'Cartier without oil or adjustment' or 'high-performance Cartier under vacuum with a 32-day power reserve', but these long-winded descriptions would only partially convey their innovative character. In 2009 Cartier produced the ID One, an unexpected prototype. This mechanical watch does not require the slightest adjustment, and its tourbillon movement does not need watch oil to reduce the inevitable wearing of its components through friction. As watchmakers well know, the oil that lubricates the cogs of a watch interferes with the working of the mechanism and makes it lose some of its accuracy. Instead of trying to get around this difficulty, or to perfect what already exists, Cartier took a radical decision: when you can neither improve a material nor change it, you simply remove it. The ID One thus aims to be the first watch that does not need adjustment and that works almost forever. To ensure the watch could withstand the absence of lubrication, its mechanism, contained in a 'Ballon Bleu' casing, was completely rethought, as was the choice of materials: carbon crystal, niobium-titanium, Zerodur and so on.

Opposite: A bird's-eye view
of the ultra-sophisticated
dial of the Cartier ID Two
watch; its casing was fitted
with the watch's components
in a vacuum.

Below: The movement
of the Cartier ID Two
and its self-lubricating
components, which are
impervious to wear, thanks
to their black coating of
ADLC (amorphous
diamond-like carbon).

After introducing the innovations in ID One, Cartier's researchers pushed the boundaries of watchmaking further with the ID Two model. Their goal was to improve the movement's performance by increasing stored energy by about a third and halving energy consumption. It is worth remembering that almost two-thirds of the energy supplied by a conventional mechanical watch's spring is lost, chiefly through friction and air resistance. The solution is to eliminate all friction, to place the casing in a vacuum. 'To shield the oscillator from air resistance as it beats, the interior of the single-block case/glass in Ceramyst transparent ceramic is a vacuum. Airtightness is reinforced by using gaskets doped with nanoparticles.'

BY ABOLISHING ADJUSTMENT, CARTIER LAUNCHED A NEW ERA OF WATCHMAKING

Moreover, the springs supplying energy to the movement are made not from metal but from micro-fibreglass, making it possible to increase the power delivered. The movement's components have a black coating of amorphous diamond-like carbon (ADLC), which is self-lubricating and almost impervious to wear. In 2012 Bernard Fornas, then the CEO of Cartier International, declared that he was convinced that the technological solutions produced by ID Two would be found in all Cartier watches of 2020 or 2030. This is already true of the Astrotourbillon Cristal de Carbone (about £176,000 or $226,000), which incorporates some of the innovations of the ID One and ID Two watches.

ROGER DUBUIS EXCALIBUR QUATUOR

FOUR SPIRAL BALANCES FOR AN INCREDIBLE WATCH

Not one, or two, or three – but four! In 2013 watchmaker Roger Dubuis produced an outlandish watch with four silicium spiral balances that corrected the effects of the earth's gravity. A watch is constantly subjected to changes in speed of operation, because its position is constantly altering with the movements of the wearer's wrist. A tourbillon – an invention patented by Abraham-Louis Breguet in 1801 – generally takes a minute to counteract these differences; but this piece, produced in an extremely limited edition, does it immediately, due to the position of its silicium spiral balances at 45 degrees and their interaction via five differentials. The Excalibur Quatuor's movement, which comprises 590 components and was awarded the Poinçon de Genève, runs at very high frequency (16 Hertz) and features a power reserve display (40 hours) of unequalled accuracy.
The four balances, each of which make four alternations a second, ensure that the Excalibur Quatuor's capacity for oscillation is augmented and that all its balances do not oscillate simultaneously. It is a new milestone of absolute accuracy – and has a new sound. The traditional ticking is replaced by a melody that brings to mind a chorus of crickets. Impressive – and worth around £300,000 ($385,000).

It took seven years to perfect this profoundly original piece. In a 48mm-diameter casing, the four spiral balances of the Excalibur Quatuor instantly compensate for the changes in operation speed caused by the position of the wrist. Whereas a conventional watch with a balance oscillating at 4 Hertz is considered to be extremely accurate, the Excalibur Quatuor pushes the boundaries of high frequency at 16 Hertz. The following pages show the iconic movement of the Excalibur Quatuor and its 590 components, from the front and the back.

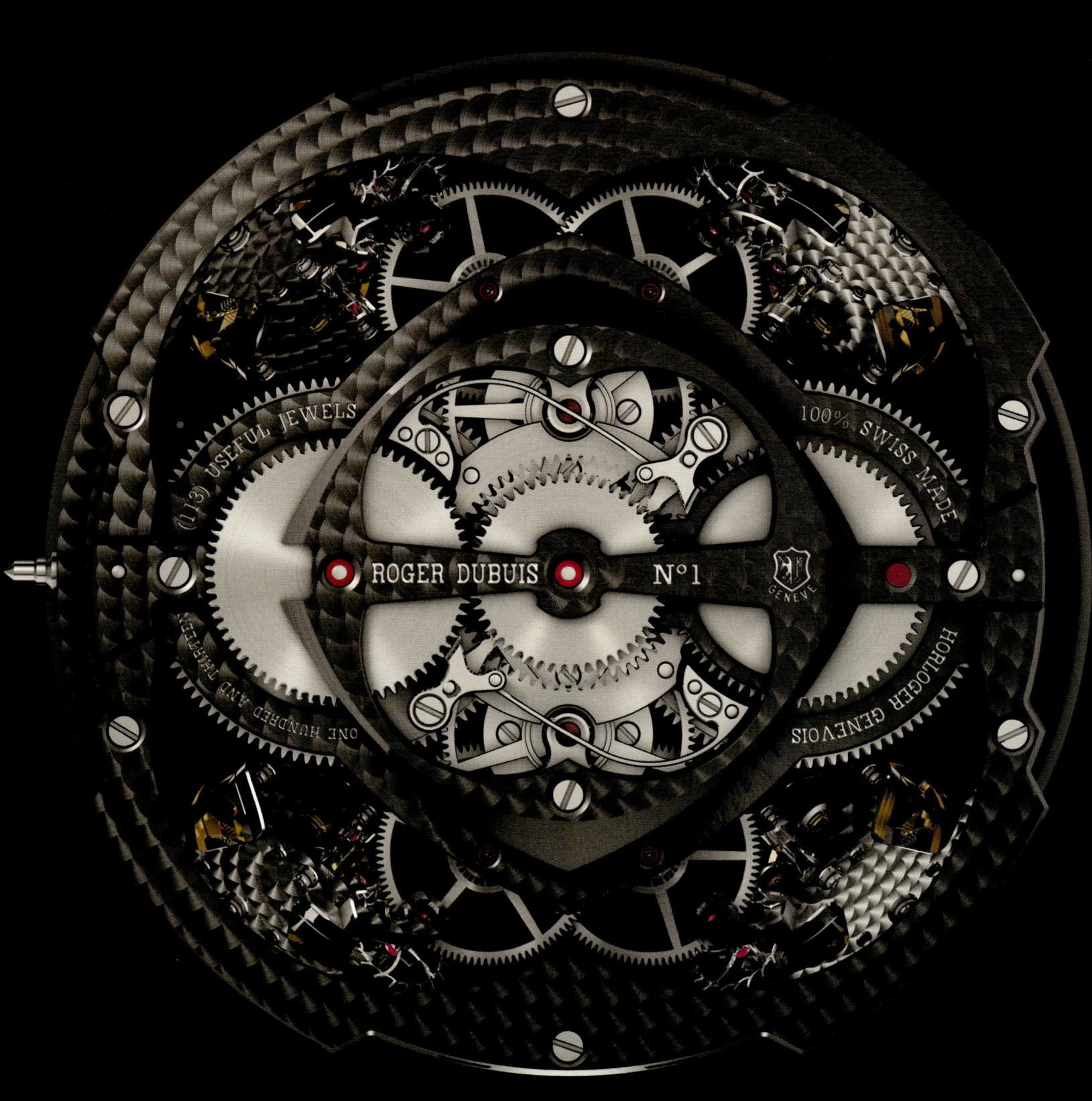

AUDEMARS PIGUET ROYAL OAK

OFFSHORE GRANDE COMPLICATION

HIGH-LEVEL SPORTS WATCHMAKING

Launched in the early 1990s, the Royal Oak Offshore is the luxury sports watch par excellence. In 2013 the Le Brassus watchmaker fitted it with an automatic mechanical movement with grand complications. This piece, of which only three were made, boasts a minute-repeater, a perpetual calendar and a flyback split-second chronograph. These functions, which normally appear on conventional watches, take on a completely different appearance on a pumped-up sporting version, clearly embodying the brand's motto: 'To break the rules, you must first master them.'

Audemars Piguet's Royal Oak Offshore Grande Complication sums up the watchmaker's art. Its movement (calibre 2885) comprises 648 components, contained in 8cm³ (½in³) ceramic and titanium casing, 44mm in diameter. It features three impressive functions: a minute-repeater, a perpetual calendar and a flyback chronograph. The perfection of each piece matches the price of this work of art, which exceeds £600,000 ($770,000).

LOUIS MOINET METEORIS

A SPACE ODYSSEY

The Frenchman Louis Moinet (1768–1853) was one of the most talented watchmakers of his day; he was one of the inventors of the chronograph whose history continues to inspire. Today, his name is borne by an independent watchmaker with no lack of ambition, as the Meteoris collection demonstrates: four watches with a tourbillon, their dials carved out of a meteorite, housed in a masterly planetarium 1.5m (5ft) high. Fitted with exclusive hand-wound tourbillon movements, the Mars and Asteroid models boast a grey gold casing set with baguette-cut diamonds; the Rosetta Stone and Moon versions feature a rose gold casing. The price? Precisely £3.8 million ($4.9 millon) to acquire this unique quartet.

METEORIS ROSETTA STONE (ABOVE)

The meteorite used to make the dial of this watch is the oldest known rock in the solar system and the oldest meteorite found on our planet. The complex rose gold casing comprises 50 components. The meteorite (Sahara 99555) was found in the Sahara Desert in 1999 (*Meteoritical Bulletin*, No. 84, 2000).

METEORIS MARS (LEFT)

A meteorite from the planet Mars is embedded in the dial of this watch, which is hand-engraved and decorated to imitate an astrolabe. Never before had a stone from Mars been used in watchmaking. This one is from a meteorite (Jiddat al-Harasis 479) discovered in Oman in 2008 (*Meteoritical Bulletin*, No. 97, 2009).

METEORIS MOON (ABOVE)

The Meteoris Moon watch is the first tourbillon featuring an authentic lunar meteorite,
recognizable by its dark colour and fine occlusions. This moon stone (Dhofar 459)
was discovered in 2001 in Oman (*Meteoritical Bulletin*, No. 89, 2005).

METEORIS ASTEROID (RIGHT)

The meteorite used in the hand-engraved dial of this watch comes from Italy,
a mysterious asteroid that was formed close to the sun and that has enabled scientists
to gather precious information about the formation of the solar system. The meteorite
was found in Western Sahara in 1990 (*Meteoritical Bulletin*, No. 85, 2001).

CHANEL J12 RÉTROGRADE MYSTÉRIEUSE

■

WINDING TIME VERTICALLY

In 2010, to celebrate the 10th anniversary of the J12 watch, Chanel brought out the J12 Rétrograde Mystérieuse. It is a perfectly round ceramic watch, with a vertically retractable crown that disrupts the progress of the minute hand. It is an absurd complication, and therefore outstandingly beautiful and highly sought-after.

In the world of business or the military, this would be called a demonstration of strength. In that of high-level watchmaking, it would, rather, be described as a demonstration of grace. For the 10th anniversary of its iconic watch, the J12, Chanel wanted to rethink the complication by going off the beaten track. The challenge was to create a perfectly round model. For this, Chanel turned to the teams of Giulio Papi, one of the great masters of high-level Swiss watchmaking. How could they create a watch that was both complicated and totally round? How could they combine Chanel's artistic heritage with the Swiss watchmaking tradition of minute mechanical work? Giulio Papi replied: 'You need to move the crown!' Why not locate it on the dial? This raised the question of the hands: how could they go round the dial with the crown in their way? This in turn raised a second problem: the crown would protrude from the watch's surface. So why not have a retractable crown? Ultimately, it was necessary to design, in a tourbillon movement, a minute hand that could not only go round the dial, but also round the crown located on that dial. As can be imagined, this was not a simple matter.

The only way to achieve this was to have the minute hand advance until it bumped into the crown, then make it go backward until it could reposition itself on the other side of the crown. But if it went back after having advanced, what would happen to the watch's accuracy? And how could it display the time during that retrograde interval? The solution was this: when the hand could no longer display the minutes, a digital display would have to take over. This ingenious idea in no way interfered with the movement of the hour hand.

The Chanel J12 Rétrograde
Mystérieuse watch, with
casing in black or white
ceramic and white or rose
gold, hand-wound
mechanical movement
designed exclusively for
Chanel by the watchmaker
Renaud et Papi (APRP SA),
ceramic plate, a retrograde
minute hand, tourbillon
movement, a ten-day power
reserve, a retractable crown
and a bracelet in black
or white ceramic.

Implementing this iconoclastic concept was clearly a very complex task. In order not to bump into the retractable crown located in the centre of the dial, the minute hand reverses its course for ten minutes: these minutes, having become invisible, are then displayed in a magnifying window in the lower part of the dial. This is a first in the history of watchmaking, as is the vertical crown integrated in the dial, which is deactivated by pushing it flush with the glass, and reactivated by finger pressure, making it rise up again to be operated.

In concrete terms, during the first ten minutes, hours and minutes are displayed in a very conventional way, in the centre of the watch. At the tenth minute of each hour (2 o'clock on the dial), the minute hand sets off in reverse, going backward and taking ten minutes to return to its standard position at the twentieth minute (4 o'clock on the dial). During this reverse revolution of 300 degrees, the minute hand completes a retrograd course of 300 degrees at a pace of 6 degrees per minute, it thus 'regresses'

A MINUTE HAND THAT GOES ROUND THE DIAL AND ROUND THE CROWN

at the rate of five minutes on the dial per minute. During the ten minutes of its retrograde movement, elapsed minutes can be read on a disc engraved with the figures 11 to 19, which moves slowly while the minute hand sets off in reverse. It does not display any figure for the 50 minutes during which the minute hand rotates normally.

After the ten minutes of retrograde movement, the time reverts to being read in the normal way. Even more impressively, at 11 o'clock a power reserve hand displays the movement's running time until it is next rewound by hand. It is designed to run for ten days once the two parallel barrels have been completely rewound.

Protected by a patent filed by Chanel, the J12 Rétrograde Mystérieuse stands out by virtue of three unconventional innovations. The first is its vertical retractable system, making it possible for the crown to disappear into the casing simply by finger pressure. The second is the construction of a perpendicular connection between the movement and the crown's shaft. The last is the creation of a mechanism comprising two ceramic buttons set into the bezel, which make it possible to control the function of the crown: button no. 1, at 4 o'clock, activates the rewinding of the movement; button no. 2, at 2 o'clock, activates setting the time. This is, in the strictest sense of the term, a triple complication: tourbillon, retrograde minute hand and retractable crown. A combination that is also a world first.

BREGUET NO. 160 POCKET WATCH 'MARIE-ANTOINETTE'

■

FOR A CERTAIN MARIE-ANTOINETTE

Commissioned in 1783 for Marie-Antoinette, the Breguet no. 160 pocket watch was not complete until 1827: neither the Queen nor its creator, Abraham-Louis Breguet, lived to see it finished. After passing through the hands of various collectors, it was stolen from Jerusalem's Museum for Islamic Art in 1983, then returned as if by magic in 2007. An extraordinary fate for a unique piece of high-level watchmaking, emblematic of Breguet's expertise.

Paris, 1783. In the workshop at 39 Quai de l'Horloge, Abraham-Louis Breguet had just received a mysterious commission: one of the officers of Marie-Antoinette's personal guard wished to make a gift to the Queen of France of the most spectacular watch possible, featuring all the watchmaking expertise of the time. Wherever possible, gold should replace other metals; complications were to be multiple and varied. There was no limit to the time or money involved. Breguet, already a supplier of the royal court, was given carte blanche.

Abraham-Louis Breguet, a Swiss watchmaker born in Neuchâtel, did his basic training at Versailles, where he rubbed shoulders with Ferdinand Berthoud and Antoine Lépine, the then masters of the discipline. Breguet was far-sighted. He founded his business in 1775 and made his mark on history with many inventions: gong springs, escapements of all kinds, the use of rubies, the tourbillon mechanism and so on. 'The admirer may have been the Swedish Count Axel de Fersen, Officer of the Queen's Guards, who according to some was her lover and according to others merely a confidant,' states a detailed investigation published in the Swiss daily Le Temps on 19 December 2007. 'The person placing the order may also have been Marie-Antoinette herself, who wanted, in complete secrecy, to present an extraordinary gift to her spouse, who was passionate about high-precision machinery.'

In 1783 Breguet was not entirely sure who his client was, but this did not matter: he had been working independently for less than ten years and wanted the watch destined for the Queen to be a masterpiece. It would therefore be 'perpetual' – that is, automatic (a feature that he was almost alone in mastering at the time) – with an oscillating mass made of platinum that had a special geometry. The watch would feature all possible complications: a minute-repeater, a complete perpetual calendar displaying the day, date and month, the equation of time, a power reserve indicator, a metallic thermometer, a large independent second hand (which made this watch a forerunner to the chronograph), a small direct-drive second hand, a pallet escapement, a gold balance spring and double *pare-chute* shock-protection device. All the movement's friction points, sinks and bearings are in sapphire, and the casing is in gold. The watch also features two dials: the first in white enamel and the second in rock crystal. The superimposition and synchronization of the various complications in these dials were, for the time, extraordinary achievements.

In 2005 Nicolas G Hayek (who in 1999 incorporated the Breguet brand into his watchmaking empire the Swatch Group) was determined to make an identical reproduction of the 'Marie-Antoinette' watch. This was the Breguet pocket watch Marie-Antoinette Grande Complication no. 1160 (opposite, far right). The Queen's favourite tree, an oak in the park of Versailles, happened to need felling at that time, so Hayek decided to give it a new life by carving the watch's case (left) from its wood. Versailles gave the tree to Breguet as a gift; as a sign of its gratitude, the company financed the restoration of the Petit Trianon. In 2007, just as the watch's manufacture was reaching its conclusion, the loot from the 1983 heist suddenly reappeared in Jerusalem – a coincidence that raised a smile at Breguet.

Opposite, top and bottom left: The front and back of the historic watch.

The challenge was immense, but nonetheless within Breguet's capabilities. In his company's registers it is described as a 'gold watch' and 'perpetual minute-repeater, perpetual equation of time, seconds at a glance'. Then came the revolution. In 1793, helped by his friend Marat, Breguet emigrated to Switzerland. He took with him the unfinished watch no. 160 – a wise precaution, for his Paris workshops were wrecked by the revolutionaries. Breguet returned to Paris in 1795 and went back to work in 1809, but the design of this masterpiece demanded too much time, and he was busy coming up with new inventions and encountering great commercial success. By 1814 the watch was almost finished – but Breguet was appointed watchmaker to the navy and was a member of the Bureau des Longitudes. He barely had time to think about Marie-Antoinette's watch.

In August 1823 Breguet turned once again to watch no. 160, which was almost complete – but he died a few weeks later. It was finished only in 1827 by the watchmakers employed in his workshops, under the direction of his son Antoine-Louis Breguet.

NO LIMIT TO TIME OR MONEY TO MEET THE ROYAL CHALLENGE

The original specification had been adhered to beyond expectations: this unbelievable watch – its cost had risen to 17,000 gold francs: a real fortune – was to hold the title of the world's most complex watch for almost a century.

So began the eventful life of the 'Marie-Antoinette'. In 1838 the Marquis of La Groye, who had been page to the Queen, took it to the Breguet workshops for a service. He was extremely elderly and never returned to collect it. The famous 'gold watch' lay dormant in a drawer at the Breguet company until 1887; it was then sold to a British collector, Sir Spencer Brunton. After that, it changed hands several times before being bought by Sir David Salomons, London's first Jewish mayor. When he died, in 1925, his daughter Vera inherited the 'Marie-Antoinette' along with other major pieces from her father's watch collection. She was a philanthropist, and founded a museum in Jerusalem devoted to Islamic art, which bears the name of her art teacher, L. A. Mayer. Here, Vera Francis Salomons placed her father's priceless watch collection. On Saturday 16 April 1983, in the middle of the night, the museum was burgled: 40 watches, including the 'Marie-Antoinette', were taken. It was found again 24 years later, on 14 November 2007, after an interminable investigation by Interpol, ransom demands and fruitless negotiations between lawyers acting as intermediaries. 'We now know that the burglary in April 1983 was the work of Naaman Diller, known as Lidor, one of the most notorious Israeli villains of the time, who died in 2004,' explains Emmanuel Breguet in *Breguet: an Apogee of European Watchmaking* (2009, Musée du Louvre Éditions and Somogy éditions d'Art). 'On his deathbed, he told his wife the story of the operation and entrusted her with a secret: the loot, which was far too well-known to be sold, was still lying in safe deposit boxes in Europe and the United States. Naaman Diller's widow then took steps in secret which culminated in the watch's return in 2007.' A bizarre adventure with a happy ending.

PARMIGIANI FLEURIER FIBONACCI WATCH

■

THE WATCH FIBONACCI WOULD HAVE LIKED TO DESIGN

The first rough plans for the Fibonacci date from 1996; they were the work of the most highly qualified watchmakers at the Parmigiani Fleurier company. An exceptional pocket watch in grey gold with more than 50 carats of diamonds was completed in 2010 and sold to a Korean collector for almost £1.5 million ($1.9 million).

A classic grand complication, with its minute-repeater, perpetual calendar and moon phase indicator, this unique timepiece pays homage to Leonardo of Pisa, known as Fibonacci, one of the greatest mathematicians of the Middle Ages, to whom we are indebted for the introduction of Arabic numerals to Europe, as well for the famous Fibonacci sequence and the golden ratio ϕ 1.618. The theme of the waterlily is subtly interpreted here by a composition of semiprecious stones cut to size. Spread out over a leaf and floating on the water, the petals poetically yet precisely suggest Fibonacci's numerical spiral. The chain's links, connected to the casing by a precious swivel, are in grey gold and set with 2,309 brilliant-cut diamond, 12 emeralds, 14 rubies and a sapphire.

The Parmigiani Fibonacci watch has a high-level movement with an exceptional finish; it also draws on the prestigious expertise of several artistic crafts, such as enamelling, setting precious stones and engraving. The dial is a fine composition of semiprecious stones: rhodonite for the waterlily's flower, nephrite jade for the leaves, black onyx for the water, and pink or green mother-of-pearl for the counters. Brilliant-cut diamonds depicting water droplets are placed on some of the leaves, making it easier to read the hours and minutes. The two backs, on the side of the movement and the side of the dial, are decorated with the same waterlily, and are mirror images of one another when the two cuvettes are open. Rendered in translucent *grand feu* enamel and set with precious stones, the motif possesses unique brilliance and depth, enhanced by the 288 trapeze-cut diamonds that adorn the rim of the two bezels.

PORTFOLIO

■

HARRY WINSTON: THE GREAT OPUS ADVENTURE

In 2001 Harry Winston brought out the Opus series of watches. Every year the New York company gives a watchmaker the opportunity to create the watch of their dreams by exploring the *terra incognita* of high-level watchmaking. Everything becomes possible – genius joining forces with expertise, artistic creativity with technical exploration. Each watch of the Opus series has been designed by a different independent watchmaker. Few brands have been up to taking the risk of total innovation.

OPUS ELEVEN, DENIS GIGUET, 2011

Created by the engineer Denis Giguet, this incredible timepiece consists of three cylindrical shapes on three levels, configured in such a way as to 'deconstruct time'. The hour is displayed by an unprecedented system of 24 placards via a system that regulates and manages the speed of the mechanism. Every 60 minutes, a mechanical ballet takes place amid visual chaos to display the hour in the centre. The minutes can be read digitally on the second cylinder, located at 2 o'clock, due to two discs, the first a semi-jumping disc for the tenths of a second, and the second a running disc for whole seconds. The third and lowest cylinder reveals a large balance in titanium. Each of the 111 examples of this limited edition features a white gold casing.

OPUS ONE, FRANÇOIS-PAUL JOURNE, 2001

'Every time I create a movement, I strive to be as honest as possible; I look for scientific truth,' explained François-Paul Journe, who founded the eponymous brand in 1999. 'No matter how delicate and complex a mechanism is, it must above all be reliable and functional. Strangely, my chronometers have been compared to the point of a diamond, probably because they convey a certain purity. I have noticed that the search for truth is something that the Harry Winston company understands perfectly.' François-Paul Journe made the first Harry Winston Opus watch, which is divided into three different movements. The 'resonance chronometer' works via the physical phenomenon of resonance. It features a mechanical movement with two gear trains, two escapements, two independent balance wheels and two indicators (hours, minutes, seconds) that make it possible to set two time zones. The platinum casing is set with 51 baguette-cut diamonds (4.85 carats), and the turquoise dial with 116 diamonds and 23 baguette-cut diamonds (0.35 carat).

OPUS 3, VIANNEY HALTER, 2003

A radical creative who enjoys nothing more than shaking convention, Vianney Halter agreed to make the third watch in the Opus series. After two years of work (much more, according to some), he was able to produce a mechanical timepiece featuring a completely new way of telling the time. The Opus 3 is, in fact, the first mechanical watch that displays the hour digitally, breaking from the rules of classic watchmaking and freeing itself from the sacrosanct hands. Here, six windows display the hour, the minutes and the date; the last four seconds of a minute are counted down on a display in the upper left window. The hand-wound mechanical movement in German silver, featuring a disc display without hands, is by Vianney Halter. In place of the traditional vertical winding crown, the Opus 3 watch has a horizontal crown that moves from high to low according to four setting positions. The casing, consisting of 25 pieces, is in pink gold, with a further 25 platinum pieces and 5 platinum pieces set with baguette-cut diamonds and round diamonds (4.44 carats).

OPUS V, FELIX BAUMGARTNER, 2005

This piece is immediately recognizable as the work of that genius of Swiss watchmaking, the founder of the Urwerk brand. Here, three small cubes – each with four figures, arranged satellite-fashion around a three-dimensional rotating system that pivots on two axes – display the time and 'point' to the minutes with a retrograde hand of great technical complexity on a scale graduated at 120 degrees. The crown is concealed by a click-spring crown protector that opens upward, like the gullwing door of a racing car. The Opus V also features, on the back of the casing, a service indicator graduated from 0 to 5 years, which warns its owner when a service is required. It was produced in a limited edition of 100 pieces: 45 in pink gold, 45 in platinum, 7 in platinum set with 950 diamonds, and 3 in platinum set with baguette-cut diamonds. The mechanical movement is hand-wound.

OPUS 8, FRÉDÉRIC GARINAUD, 2008

This watch is probably one of the most singular of the Opus series. Inspired
by the 1970s, with horizontal casing resembling a TV, it embodies, according to
Harry Winston's official terminology, 'the art of digital emotion'. Its hand-wound
mechanical movement displays the hours digitally via a bolt located on the right-hand
side of the casing – the winding crown being on the left. The mechanism thus links
together all the functions: minutes, hours and the AM/PM indicator, so that everything
can be displayed on demand, as on LED watches of the 1970s. The design of the
white gold casing echoes that of the dial, which resembles a display module,
with segments in shot-blasted steel and sides in amorphous carbon.
The watch was a limited edition of 50.

CHOPARD CHOPARDISSIMO WATCH

■

A FLOWER-WATCH ADORNED WITH DIAMONDS

In 1997, at the Basel Watch Fair, Chopard created a stir by presenting a lady's watch of astounding beauty. It is an exceptional piece, featuring three heart-cut diamonds in different colours (with a total of 38 carats) resting on a bracelet consisting entirely of yellow and white diamonds. The three diamond hearts can open out like the corolla of a flower, revealing a watch covered with yellow diamonds. An ingenious mechanical system allows them to rotate, giving a glimpse of the dial of this watch, which has a quartz movement. Requiring 2,000 hours' work to make, the Chopardissimo was a $25 million (£19.5 million) masterpiece – at the time, the world's most expensive jewellery watch.

An ultra-feminine watch, the Chopardissimo in white, rose and yellow gold is set with three heart-cut diamonds: one pink (15.37 carats), one blue (12.79 carats) and one white (11.36 carats). It is also decorated with 491 yellow brilliant-cut diamonds (13.76 carats), 260 white pear-shaped diamonds (73.77 carats), 91 white brilliant-cut diamonds (10.29 carats) and 29 yellow pear-shaped diamonds (25.52 carats).

GRAFF HALLUCINATION

■

THE WATCH THAT SHINES WITH A THOUSAND LIGHTS

Revealed in 2014, the aptly named Hallucination watch by the British jeweller Graff Diamonds resembles no other. A kaleidoscope of coloured diamonds of the greatest rarity, this jewel that tells the time is considered to be the world's most expensive quartz watch. It required thousands of hours' work and its price speaks for itself: $55 million (£42.9 million).

Yellow, blue, pink, green, orange and silver, the diamonds of this Graff watch were chosen, cut and set by a team of designers, jewellers, gemologists and experienced master artisans. Graff Diamonds, founded in 1960, specializes in the cutting and sublimation of exceptional gemstones, such as the Wittelsbach-Graff, a 31.06-carat blue diamond bought in 2008 for £16.4 million ($21 million) and considered the most expensive diamond ever sold.

GRAFF FASCINATION

■

THE FASCINATION OF BEAUTY

This lady's watch is set with 152.96 carats of diamonds that are among the purest in the world. Revealed in 2015 by the British jeweller Graff Diamonds at the Basel Watch Fair, it is also convertible: the 38.13-carat pear-shaped diamond of D Flawless quality is removable and can be used to adorn a ring or a bracelet. Its price – $40 million (£31.16 million) – makes it unique; it embodies all Graff's historical experience of making exceptional jewellery and convertible watches with mechanisms carefully designed to be invisible.

'We are inspired by the rarest and most precious diamonds – their influence is evident throughout everything we do,' explained Laurence Graff, founder and chairman of Graff Diamonds. 'The Fascination is an outstanding piece, carefully crafted so it can be worn in a number of different ways – adding a touch of magic to the jewel.'

JACOB & CO. BILLIONAIRE

■

FOR BILLIONAIRES ONLY!

Famous for its outlandish creations, Jacob & Co., the unconventional company founded by Jacob Arabo
in 1986, flaunts its 'bling'. It silenced all jealous voices when it revealed the Billionaire in 2015.
This unique $18 million (£14 million) lady's Billionaire watch boasts 260 carats of superior-quality
diamonds, as well as a mechanical skeleton movement fitted with an exclusive tourbillon. The setting
of the gems is of exceptional quality; to ensure their settings are invisible, each diamond is fitted
in an inverted pyramid of white gold pearls.

This watch is the result of two years' work. All the gems had to be of the same quality
and have the same shape, depth and polishing; certain reductions had to be conceded
to achieve this uniformity. The gems then had to be perfectly graded, the first row
being 3 carats, the next 2.5 carats, then 2, and so on down to 0.5 carat on the buckle.
Everything had to match and be invisible; the joins between the gems are 'seamless'.
Last but not least, the winding crown is a 2-carat diamond.

VACHERON CONSTANTIN KALLISTA

■

KALLISTA, THE FAIREST OF THEM ALL

In 1979 Vacheron Constantin brought out a dazzling gentleman's watch, sculpted directly from a gold ingot and set with 130 carats of emerald-cut diamonds: it is called Kallista, Greek for 'the most beautiful'. This unique model, from a design by Raymond Moretti. was at the time the world's most expensive watch. It was the result of more than 6,000 hours of painstaking work, combining watchmaking expertise with a jeweller's skill. It is decorated with 118 diamonds, each weighing between 1.2 and 4 carats. And that is not all: its mechanical movement (calibre 1052) was considered at the time of manufacture to be the world's thinnest in its category.

The casing, dial and bracelet of the Kallista watch are in yellow gold. Its back
displays the profiles of a man and woman, as if this timepiece were dedicated to love.
Thirty years later, in 2009, in homage to the Kallista, Vacheron Constantin produced
the Kallania, its heir: 186 emerald-cut diamonds (170 carats) and the world's finest
mechanical movement (calibre 1003), bearing the Poinçon de Genève.

CARTIER SECRET WATCH WITH PHOENIX DECORATION

■

CARTIER'S FABULOUS MENAGERIE OF A WATCH

Barbara Hutton's tigers, María Félix's snake, the Duchess of Windsor's panther: Cartier is not just the king of jewellers and the jeweller of kings – the company in Paris's rue de la Paix also excels in the art of taming the wildest animals into sumptuous jewels. A 'secret' watch made of more than 3,000 diamonds, this unique piece, which pays homage to the majesty of the phoenix, embodies all Cartier's expertise in the world of high-level jewellery watchmaking.

Every year Cartier goes hunting for the most beautiful gems in the word. It then sets them, as no other company knows how, on unique jewellery watches that reinterpret its wild fantasy world. Cartier's Menagerie brings together precious panthers and tigers, glittering snakes and eagles – and above all keeps up a tradition that allows the company, year after year, to prove its unquestionable supremacy. Combining realism and storytelling, the Parisian jeweller unites jewel and watch. Made in 2010, the 'secret' watch with phoenix decoration is part of this tradition. This unique piece required more than 2,500 hours' work. On a white gold structure set with 3,010 diamonds (80.113 carats), a 3.53-carat pear-shaped diamond reigns supreme, displaying curves of exceptional refinement. Here, the time display is incorporated into the decoration and establishes a dialogue with the mythical bird. A breathtaking watch for $2.7 million (£2.1 million): the price of a dream.

PIAGET LIMELIGHT EXCEPTIONAL PART REFERENCE GOA34135

■

ONE WATCH, TWO DIALS, 506 DIAMONDS

'Always do better than necessary.' The motto of Piaget, the extraordinary creations of which combine top watchmaking expertise with the sophistication of high-level jewellery, is perfectly embodied in this piece of pure visual delight.

In 2009 Piaget revealed its Limelight collection. Among its pieces was an exceptional 'secret' watch that re-examines nature in a refined way: hundreds of diamonds cut in different ways make up this rare, shining work of art.

An invisible, almost magic, opening mechanism allows two leaves – each made from a fabulous marquise-cut diamond – to rise up: one unveils a dial in Polynesian mother-of-pearl, the other a dial set with 75 brilliant-cut diamonds. In total this watch is set with 122 brilliant-cut diamonds, 33 pear-shaped diamonds, 32 baguette-cut diamonds and two 2 marquise-cut diamonds. The bracelet bears 98 brilliant-cut diamonds, 109 pear-shaped diamonds and 34 baguette-cut diamonds. Piaget 56P quartz movements drive the hands on the two dials.

This marvel of jewellery, with a total of 76.2 carats of diamonds, required 1,050 hours' development work and some 160 hours' setting work. Piaget the jeweller and Piaget the watchmaker honour the same tradition, tinged with a boldness and freedom that allows creations to be constantly renewed. 'The company has forged itself a reputation that is impossible to overlook in the world of luxury goods, designing sumptuous jewels where illuminated time is blessed with the honour of eternity,' Piaget says. A bit lyrical, certainly – but undeniably true.

DIOR VIII GRAND BAL PIÈCE UNIQUE 'ENVOL'

■

DIOR'S WATCHES GO TO THE BALL

Since 2011 the Dior VIII Grand Bal collection has aimed to pay tribute to Christian Dior's party spirit, combining the essential with the frivolous. Inspired by the fashion designer's unconventional dresses, the Grand Bal collection combines the best mechanical expertise with a strong attention to detail. These models are either unique pieces or very limited editions. They are all fitted with 'Dior Inversé' ('inverted') calibre automatic movements, where the oscillating mass, both functional and variously decorated – with woven gold thread, feather or mother-of-pearl inlay, gold netting or diamonds – is revealed on the dial, creating an effect as magical as it is mysterious. These are certainly high-fashion watches.

The extraordinary dial of the Dior VIII Grand Bal Pièce Unique 'Envol' no. 14 watch consists of inlaid work featuring the wing cases of blue scarabs, while the faceted hour and minute hands are hand-painted in fluorescent pink. The casing is in white gold, the bezel tiled with two rows of baguette-cut sapphires, the rim of the bezel set with brilliant-cut diamonds and the crown set with a rose-cut diamond. This is a formal watch of outstanding beauty. The automatic mechanical movement features a 'Dior Inversé 11½' calibre, which displays a functional oscillating mass in the dial. This mass, in white gold, is cut out and decorated with mother-of-pearl and baguette-cut diamonds. It is a masterpiece of technical prowess and sophistication that deserves its patent. The following pages show some variations of the Dior VIII Grand Bal watches.

BLANCPAIN LE BRASSUS EROTIC WATCH
WITH CARROUSEL AND MINUTE-REPEATER

■

THE EROTICISM
OF TIME

Erotic watches, which first appeared at the end of the 17th century, are decorated with risqué scenes –
animated or otherwise – often concealed by clever mechanisms. Highly sought-after by major collectors,
they are made to order, meaning most are unique. Many erotic watches feature a minute-repeater, which
makes it possible to trigger small automated figures that indulge in various frolics. They were historically
banned by the religious authorities, and so were made in the utmost secrecy – as they still are today.
Since the launch of its calibre 232 (the world's first wristwatch with a minute-repeater with automated
figures) in 1993, Blancpain, along with Ulysse Nardin, Svend Andersen and
Girard-Perregaux, has been one of the most active Swiss watchmakers in this field.

This is an erotic watch by Blancpain, featuring that maker's hand-wound calibre 232
with carrousel, minute-repeater and cathedral chime. The dial is in *grand feu* enamel.
The back of the casing, in red gold, depicts an erotic scene. It is a unique piece.

WATCHES
FOR HEROES

OMEGA CALIBRE 540 'JOHN F. KENNEDY'

■

JFK'S PROPHETIC OMEGA

In 1960 John Fitzgerald Kennedy was the Democratic candidate in the US presidential election. That summer Edward Grant Stockdale, a businessman and deputy in the Florida legislature, gave him an elegant rectangular Omega watch in yellow gold. On the back of the casing he had the prophetic words engraved: 'President of the United States John F. Kennedy from his friend Grant'.

It was a bold gift; when JFK received this watch, the presidential election had not yet taken place. It was not until 8 November 1960 that John F Kennedy narrowly beat Richard Nixon, thus becoming the youngest head of state in American history at the age of just 43. Featuring an extra-thin mechanical movement (2mm thick), this Omega watch from his faithful friend seems to have brought him luck.

An Omega watch in yellow gold, with Omega's extra-thin mechanical calibre no. 540. This model, with its silver dial and two-piece rectangular casing with lateral bars, was launched in 1957 and exists only in an 18-carat gold version. JFK wore it the day of his inauguration ceremony, 20 January 1961 (previous page).

This timeless, stylish Omega wristwatch, which the American president always wore, was for a long time in the Robert White private collection, devoted to the life and career of JFK. It was bought back by the Omega Museum in December 2005 for $350,000 (£272,650), at a Guernsey's auction held in New York. It was offered for sale with two 'thank you' letters addressed to Grant Stockdale, one from JFK, dated 4 April 1962, and the other from his wife, Jackie Kennedy, dated 23 August 1960. Legend has it that JFK,

A LUCKY WATCH – PROOF OF A FAITHFUL FRIENDSHIP

out of courtesy, never forgot to wear it when meeting his friend Grant Stockdale. In the book *Omega: A Journey Through Time* (2007), the Swiss brand's historians add that the president never failed to point out to him that he was wearing 'the Stockdale watch', as he had affectionately nicknamed it. Such was the friendship between the two men that ten days after the assassination of JFK, on 22 November 1963 in Dallas, Grant Stockdale, inconsolable, threw himself off the 13th floor of the Dupont Building in Miami, where he had his office.

RICHARD MILLE RM 27-01 AND 27-02 TOURBILLON RAFAEL NADAL

■

GAME, SET
AND WATCH

Since 2010, on all the world's tennis courts, Rafael Nadal has worn one of his Richard Mille watches on his right wrist – he is left-handed. The watchmaker made five for the Spanish champion; of these, the RM 27-01 Tourbillon is unquestionably one of the most impressive. It is a high-tech ergonomic watch weighing no more than 20g (¾oz), capable of withstanding accelerations of thousands of G-force, made from futuristic materials – and costing more than £500,000 (£640,000).

It all began in 2008, when Richard Mille met Rafael Nadal through a mutual, tennis-playing friend. But Nadal, partly out of superstition, refused to wear a watch on court. A few weeks later, during a meal with King Juan Carlos, Nadal heard the monarch speak very highly of Richard Mille, a new watchmaking company with the highest ambitions. The King's words rang true, and Rafael Nadal seems to have changed his mind. Shortly afterward, Richard Mille travelled to Majorca to meet the nine-times winner of the Roland-Garros tournament. There was an immediate spark between the two men – so much so that in 2010 Nadal trained with a watch on his right wrist, and soon decided it was time to wear one during tournaments.

This ultra-tough tourbillon watch was the RM 27. A deep casing in black carbon composite houses a unique skeleton movement in titanium and lital (an alloy containing aluminium, used notably by NASA), and weighs only 3.83g ($^1/_7$oz). Nadal wore the watch that year, when he won Roland-Garros, Wimbledon and the US Open. The tennis player was full of praise for his new watch: it was very light and extremely comfortable to wear. He even declared that if he played without the watch on his wrist, he felt like he was missing something.

In an interview with the French daily *Le Figaro* on 20 March 2015, Richard Mille related the saga of this watch: 'When he was in the city, Rafael Nadal was an admirer of our collections, but he refused to wear them during a tennis match. In order to win him over, we were obliged to design an extra-light watch weighing less than 20g (¾oz) that could withstand accelerations of more than 600*G*. When I spelt out these two goals to the development team at Renaud et Papi, they replied: "Boss, you need a holiday." It wasn't a simple task. In training, Nadal broke half a dozen watches. It took us two years to come up with the RM 27: every gram saved was a battle. Now, he can no longer play a match without his RM 27; it is his second skin.'

In 2011 Richard Mille brought out the RM 035 Chronofiable – a new, practically indestructible watch. It had a three-dimensional skeleton titanium movement, functioning on different levels and weighing 4.3g ($\frac{1}{6}$oz), which was suspended within an ultra-strong magnesium and aluminium casing. Its price ran into the hundreds of thousands of pounds. Some people invest in property, but Rafael Nadal chose to invest in this watch, with its avant-garde technology that rendered it unique.

IS ALMOST INDESTRUCTIBLE WATCH FEATURES UNIQUE TECHNOLOGY

A mountain was climbed in 2013 with the RM 27-01, the world's lightest and toughest tourbillon watch. The movement is suspended by four 0.35mm woven micro-cables, in a casing made from carbon nanotubes. The skeleton tourbillon movement, which weighs just 3.5g ($\frac{1}{8}$oz), can withstand impacts greater than 5,000G. Richard Mille and Nadal worked together on the watch's design, because it was virtually impossible to reduce its weight further. Other models followed, such as the RM 35-01, with an NTPT carbon casing, and the RM 27-02. With its cable mounting, the 2013 tourbillon model RM 27-01 – which weighs just 19g ($\frac{2}{3}$oz) – will go down in history, just like the Spanish tennis player's incredible collection of trophies.

ROLEX DATO-COMPAX REFERENCE 6236 'JEAN-CLAUDE KILLY'

■

A HIGH-SPEED HIT

Along with the Rolex Sky-Dweller, launched in 2012, and the chronographs reference 81806 of the 1950s, with their triple date displays and moon phases, the five Dato-Compax references produced between the 1940s and the early 1960s (nos. 4767, 4768, 5036, 6036 and 6236) are among the brand's most complex models. Jean-Claude Killy, a triple Olympic ski champion, was a historical ambassador for Rolex and wore several of these watches during the 1960s. Since then, collectors have named this family of hand-wound chronographs with triple date displays after him.

Jean-Claude Killy was a unique champion. Winner of three gold medals for skiing at the 1968 Grenoble Olympics, this playboy of the tracks surprised the world by retiring from the sport at the age of just 25, to embark on a career as a racing driver and actor. He took part in the 24 Hours of Le Mans in 1969 with Bob Wollek, driving an Alpine A210 and finishing 17th, won the Targa Florio race (GT trophy) in 1967, in a Porsche 911S, and appeared in several films, including George Englund's *Ski Riders* (1971). After that he had a career as a high-level manager: he was a member of the International Olympic Committee until the age of 70, a director of ASO, the company that organizes the Tour de France, the designer of numerous items of ski clothing that bore his name, and an ambassador for many brands, including General Motors, Coca Cola, United Airlines and Rolex.

Yet perhaps Jean-Claude Killy's most lasting impact was as an inspiration for Rolex, and not only because he has sat on its board of directors for more than 40 years. During the 1960s and 1970s, he appeared in numerous adverts that showcased the merits of various Rolex sporting watches. He never promoted the Dato-Compax triple date display watches, but these rare models, produced in small numbers over some 20 years between 1940 and 1960, are his favourites. Rolex collectors so often saw him wear these hand-wound chronometers during the 1960s – references 4767, 4768, 5036, 6036 and 6236, with steel, yellow gold or rose gold casing – that they called them simply the 'Jean-Claude Killy' watches, an unofficial name that Rolex has never recognized. The value of these watches is constantly increasing; the most expensive, as of today – a reference 6236 in steel, made in 1960 and bearing the number 576'392 – sold for $638,000 (£497,000) in 2012 at a Christie's auction in New York. The estimate had been between $120,000 (£93,500) and $180,000 (£140,000). A gold medal for a watch that cost only around 100 francs in the 1960s.

MAHATMA GANDHI'S ZENITH ALARM POCKET WATCH

■

THE INCREDIBLE JOURNEY OF GANDHI'S ZENITH

Stolen on a train, returned to its owner and coveted by every collector:
Gandhi's pocket watch had an unusual life.

'I may add that it had a radium disc, and also a contrivance for alarm. It was a gift to me. The cost then was over 40/-. It was a Zenith watch.' Signed by Mahatma Gandhi, the note is dated 28 May 1947. A few days earlier, during a train journey to Kanpur, Gandhi's fob watch, dating from 1910–15 and habitually worn on his belt, was stolen; it was the only possession that the advocate of non-violence and extreme poverty would wear in public. It had been given to him as a gift by Indira Gandhi, née Nehru, Prime Minister of India from 1966 to 1977 and a friend of Gandhi's.

No thicker than a conventional pocket watch, this silver timepiece has a cover hinged at 12 o'clock which, once opened, could be used as a base on a bedside table. It set the rhythm of the life of this exceptional man, who appreciated its precision. 'I hate it if I am late for prayers even by a minute,' he once said.

Opposite: An identical
replica of the Zenith alarm
pocket watch Mahatma
Gandhi carried with him.

Below right: The lot
consisting of Mahatma
Gandhi's round glasses,
bowl, plate, leather sandals
and famous Zenith alarm
pocket watch, sold
by Antiquorum in 2009.

The double-function crown winds the movement in the usual direction, left to right.
In the other direction, it supplies the power reserve of the striking mechanism, which
is set using the right-hand button. The left-hand button adjusts the time. The subdial
at 12 o'clock displays the alarm time.

'I HATE IT IF I AM LATE FOR PRAYERS EVEN BY A MINUTE.'

At the end of 1947 the thief, consumed with remorse, asked to meet with Gandhi to
return his watch and ask his forgiveness. Some time later, Gandhi gave it to his great-
niece, Abha, who had been his assistant for six years. He was to die in her arms in 1948
after being attacked by a Hindu nationalist in New Delhi. The watch then changed hands,
from one private collector to another. The adventure came to an end in 2009, when the
auction house Antiquorum offered an exceptional lot for sale, consisting of Mahatma
Gandhi's round glasses, a bowl, a plate, his leather sandals and the famous Zenith alarm
pocket watch. The collection was sold for $1.8 million (£1.4 million) to Vijay Mallya, an
Indian billionaire. It was time for this watch to go home.

TIFFANY & CO. FRANKLIN D ROOSEVELT

■

THE WATCH ROOSEVELT (POSSIBLY) WORE AT YALTA

The watches worn by US presidents have all spawned an array of literature and many myths. Featuring an annual calendar (day, date and month), Franklin Delano Roosevelt's Tiffany has a real history behind it: it is, in all probability, the watch he was wearing during the Yalta Conference in February 1945, just a few weeks before his death.

On Tuesday 30 January 1945 President Roosevelt celebrated his 63rd birthday with his family. His son-in-law, John Boettiger, handed him a handsome little pale turquoise box; inside was a watch. The most powerful man in the world, the 32nd president of the United States and the great architect of the Allies' victory against Nazism, eagerly put it on his wrist. The back of the casing bore the engraved inscription: 'Franklin Delano Roosevelt, with loyalty, respect and affection, January 30, 1945'.

This watch, which embodies the aesthetic ideals of the time, is a round model in yellow 14-carat gold made by Tiffany & Co., one of the most famous jewellers on Fifth Avenue.

Since the 1860s, man
American presidents
including Abraham
Lincoln, Woodrow Wilson
Dwight D Eisenhower
Harry S Truman and
Lyndon B Johnson, hav
been given a Tiffany & Co
watch. Roosevelt's shows
subtle touch in the way th
date is displayed, via
central hand with a red
arrow at the end, and
a circular graduation
showing the 31 days o
the month in blue

Elegant yet casual, as American watches often are, it has a Swiss Movado 10-ligne hand-wound mechanical movement with a relatively basic technical architecture: ten rubies, a simple escapement and a Breguet spiral driving an annual calendar. The dial shows a little more imagination, with its Arabic numerals, a small engraved second counter at 6 o'clock, two windows (month and day) at 9 o'clock and 3 o'clock, respectively, and the inscription 'Tiffany & Co.' duly situated at 12 o'clock.

ROOSEVELT'S TIFFANY & CO. WATCH: A PRIVILEGED WITNESS TO HISTORY

The subtlety of this model lies in the way it displays the date – via a central hand with a red arrow at the end, which points to the date via a circular graduation showing the 31 days of the month in blue. The watch bears the reference number T249743/44776. The baton-shaped hour and minute hands are in blued steel, conforming to the great Swiss tradition.

Just a few days after his birthday, President Roosevelt met with Winston Churchill and Joseph Stalin; the great winners of World War II divided up the world in the Livadia Palace, not far from Yalta, a seaside resort in Crimea. Throughout the negotiations President Roosevelt supposedly wore the Tiffany watch his son-in-law gave him. The legend is that he was also wearing it when he died, on 30 March that same year, in Warm Springs, a small Georgia spa town where the famous president, who was paraplegic liked to stay

HAMILTON VENTURA

■

A HAMILTON LEGEND FOR THE KING

Elvis Presley loved rock and roll, Cadillacs – and gold watches. He owned several, switching from one to the other as the mood took him: a square Longines, an Omega Constellation, a rare Corum Buckingham and a Rolex Cellini King Midas (reference 9630) with its characteristic asymmetrical casing. But his favourite was a Hamilton Ventura, the world's first electric watch: a pioneer of quartz watches, easily recognizable due to its avant-garde style.

From 1958 to 1960, Elvis Presley was in Germany, doing military service on the American base in Friedberg. He appears in several photographs wearing, on his left wrist, a strange, more or less triangular, watch: a Ventura, launched by the Hamilton brand on 3 January 1957, the first model driven by an electromechanical movement rather than a traditional mechanical calibre. For René Rondeau, a Californian expert and collector, the Ventura marked the first great development in watchmaking technology for almost 500 years. It was not only the forerunner of all present-day quartz watches, but also a symbol of innovation, with an incredibly futuristic style. A complete departure from other watches of its time, it was an immediate sensation. The image of 'the King' will forever be associated with this model. It personifies a triumphant America, proud of its taste and its technological supremacy. In *Blue Hawaii* (1961), one of Elvis Presley's films, a close-up shot lingers on the watch – impossible to miss.

Elvis Presley liked to wear his Ventura, with its atypical casing, on screen and in town. It had been created by Richard Arbib, one of the best industrial designers in the United States. He designed all manner of products, from vacuum cleaners for Eureka to boats for Century and cars for General Motors. For Christmas 1965, the rock star bought a second Ventura in a Memphis jewellery store, with white gold casing, a black dial and a two-tone leather strap. He was to keep it until his death, on 16 August 1977. It is now part of the collection of the Hamilton museum.

ROLEX GMT-MASTER REFERENCE 1675 AND 6542,

ROLEX SUBMARINER,

ROLEX DAY-DATE

FIDEL CASTRO, HIS CIGARS AND HIS ROLEX

Although symbols of luxury today, Rolex watches were once prized above all for their unfailing reliability and sturdiness. Fidel Castro's Rolex watches even withstood the thick smoke from his Cohiba cigars, which gave their casings a premature patina.

There is a historic photograph, famous all over the world, which shows Fidel Castro sitting at a table facing Nikita Khrushchev during a meeting in the Kremlin on 27 April 1963. There are smiles on the faces of the dozen people present, while the 'Líder Máximo' carefully lights an imposing cigar. On his left wrist are two Rolex watches: probably a GMT-Master (which allows the time to be set in two different time zones) and a Day-Date with a light-coloured dial. Fidel Castro was in the habit of wearing these two watches so that he could instantly tell the time in three time zones: Havana, Moscow and Washington. It is also said that he was wearing a Submariner when he overthrew General Fulgencio Batista in 1959, as a photograph published in the American magazine *Life* confirms.

Several sources have reported that when Castro's revolutionaries took power in Cuba, they plundered the local branch of Rolex and cheerfully helped themselves to its stock. His biographers state that Fidel Castro may have given a Rolex watch – very probably a GMT-Master reference 1675 with a 'Serpico Y Laino' lacquered dial – to his companion in arms, Che Guevara. The legend is that in October 1967, when the latter was captured by the Bolivian army, an American agent named Félix Rodríguez, covertly aided by the CIA, stole 'Che's Rolex' on his deathbed – and possibly a second one, according to certain sources – after Che Guevara had been executed and his hands amputated to identify the body. This watch has never reappeared, thus exciting the interest of collectors – as do those worn by Fidel Castro.

TAG HEUER SPORTS COUNTER CHRONOGRAPH REFERENCE 2915A

OUT TO CONQUER SPACE

The start of the conquest of space can been exactly dated: it began on 25 May 1961, when President John F Kennedy announced to Congress that he wanted to send a man to the moon – and bring him back safe and sound.

The Cold War was at its height and the Soviet Union had a head start. In fact, Yuri Gagarin had made history on 12 April 1961, when he became the first man to make a space flight, on the Vostok 1 mission. Ten months later, in February 1962, John Glenn became the first American to orbit the earth on board NASA's Mercury-Atlas 6 'Friendship 7' spacecraft. He made three orbits around the planet at a maximum altitude of 260km (162 miles) and an orbital velocity approaching 7.8km (5 miles) per second. It was a historic mission lasting about five hours, during which John Glenn did not wear a conventional watch, but a Swiss sports chronograph, a TAG Heuer Reference 2915A, then considered the best on the market and capable of measuring time to the nearest fifth of a second for 12 hours. It was attached to his space suit, over his sleeve, with elastic straps that had been made to measure. It was Jeff Stein – an American collector, independent researcher and acknowledged world expert on the Heuer brand – who made this discovery in the early 2000s. 'After the failure of other brands on the earlier flights of Alan Shepard and Virgil "Gus" Grissom, the TAG Heuer was selected by NASA due to its ability to withstand the high G-forces created by liftoff,' he notes on his respected website Onthedash, explaining that this sports counter was meant to act as a 'back-up watch' in the event of a problem. The TAG Heuer company thus became the first Swiss watchmaker to go into space.

ROLEX OYSTER PERPETUAL DATEJUST 'KONRAD ADENAUER'

THE GOLD ROLEX OF THE CHANCELLOR OF PEACE

In 1955 the Federal Republic of Germany was beginning to get back on its feet. It joined NATO and established the Bundeswehr. That year Konrad Adenauer, who had been chancellor of West Germany for six years, was given a Rolex Datejust in yellow gold.

The watch came in a leather case. Its back was engraved with the chancellor's name and his coat of arms. Very unusually, this Rolex Reference 6305/1 was accompanied by a letter signed by Hans Wilsdorf, the founder of the Geneva watchmaking company, explaining how the model functioned. It was a way for the boss of Rolex to show his admiration for the statesman behind the building of Europe and Germany's renaissance. In a memorable auction in Geneva in 2011, Sotheby's sold it for £130,600 ($167,700) – three times its estimate. The chancellor's Rolex had been carefully kept in its original state by his heirs.

OMEGA SPEEDMASTER PROFESSIONAL 'MOONWATCH' ST 105.012

THE WATCH THAT WALKED ON THE MOON

On 21 July 1969 at 2.56 GMT, when the American Neil Armstrong set foot on the moon, humanity was launched into a new era – and so was Omega. With the Moonwatch ST 105.012, a watchmaking legend was born.

In 1965, after a series of brutal tests, NASA finally chose the watch that would accompany its astronauts on their conquest of space: an Omega Speedmaster reference ST 105.003. Watched by millions of TV viewers, the commander of the Apollo 11 mission made 'one small step for man, one giant leap for mankind'. It is said that he did not wear his Omega Speedmaster on his wrist: immediately after landing, the electronic counter on board the lunar module had broken down, and Armstrong chose to leave his mechanical chronograph on board as a back-up instrument.

Previous page: The Omega
Speedmaster Professional
reference ST 105.012 watch
worn by the astronaut
Richard F Gordon for the
second moon landing in
history, that of the Apollo 12
mission (14–24 November
1969). The reference
ST 105.012 has an
asymmetrical casing and the
designation 'Professional'
on the dial. It is also the
model that Buzz Aldrin
wore on his right wrist
when on the moon's surface
on 21 July 1969.

Opposite: The American
astronaut Buzz Aldrin.

According to the same legend, on 21 July 1969 the first Speedmaster ST 105.012 actually taken onto the moon was that worn by Buzz Aldrin, the 'second man', who trod the lunar surface 15 minutes after his colleague. 'At the moment when millions of TV viewers were following the progress of the first men to walk on the moon,' the authoritative work on the brand, *Omega: A Journey Through Time* (2007), relates, 'Aldrin suddenly said: "I think my watch stopped, Neil." The people at Omega were on the verge of apoplexy! Happily, the astronaut immediately corrected himself: "No, it didn't – only

THE FIRST MOONWALKERS' WATCH

the second hand." Phew, that was close!' A few weeks later, in a letter dated 18 September 1969, the astronaut Donald Slayton threw light on the story: Buzz Aldrin had inadvertently pressed the chronometer's stop button and the 'second hand' (actually the central second hand of the chronograph function) had stopped, while the watch was still working.

Back on earth, the Apollo 11 team agreed to a request from the National Air and Space Museum in Washington, DC; all the equipment used during the mission, including Buzz Aldrin's 'Moonwatch', became part of the famous museum's collections. But the 'first watch to walk on the moon', probably one of the most famous of the 20th century, disappeared en route. It has never been found.

ROLEX SUBMARINER 5513 'Q' ROGER MOORE 007

JAMES BOND'S ROLEX

In November 2015 one of James Bond's most famous gadgets, the Rolex Submariner 5513 worn by Roger Moore in *Live and Let Die* (1973), sold for £279,600 ($359,000) at a Phillips auction in Geneva.

This 1972 watch was customized specially for its role in *Live and Let Die*, a film that has since been watched by millions. Armed with teeth as sharp as those of a circular saw, it enabled Agent 007 to slay some of his adversaries. Even more usefully, this customized Rolex Submariner contains (at least in the world of the film) a magnet that can create a magnetic field on demand – an especially useful function for protecting Bond from the bullets of those hunting him down. A great and passionate hedonist, the incorrigible James Bond – who also wears a Hamilton Pulsar P2 2900 LED watch in the movie – uses it to undress Miss Caruso, played by the English actress Madeline Smith, by unzipping her dress using the magnetic power of his Submariner 5513.

Bond's Rolex Submariner, with its inner casing signed 'Roger Moore 007', was snapped up for £167,800 ($215,500) at a Christie's auction in Geneva in November 2011. Four years later, its value had increased by almost 70 per cent. Rolex and James Bond: when two prestigious names come together, records are smashed.

BULOVA LUNAR EVA (WRIST) CHRONOGRAPH

■

BULOVA'S 'MOONWATCH'

On the Apollo 15 mission in the summer of 1971, Colonel David R Scott became the seventh man
to walk on the moon, and the only one to drive a Lunar Rover. Like all the other astronauts who
went to the moon, he wore an Omega Speedmaster Professional chosen by NASA. However, for this
mission, David Scott also took his own watch, a Bulova chronograph. Its steel casing still shows traces
of rust and moon dust. As of today, it is the only personal watch ever worn on the moon. A wealthy
Miami businessman and well-known collector of rare objects bought it for $1.625 million (£1.266 million)
at an exceptional auction held in October 2015 by the American auction house RR Auction.

On the official storage list of the personal effects the members of the Apollo 15
mission took with them in 1971, Colonel David Scott's chronograph was duly listed
by NASA under the reference SEB12100039-002. He took this watch, made by the
American company Bulova, 'just in case' – and it was just as well, for the second time
he left the spacecraft, the Hesalite crystal of his Omega became detached, fell off and
was lost in the dusty vastness of the lunar desert. Was it because of pressure, heat
or mishandling? We will never know. When he left the spacecraft for the third – and
last – time, in the massif of the Mons Hadley Delta, he decided to put his Bulova on his
left wrist; it replaced the Omega at short notice for his excursion in the Lunar Rover,
which lasted almost five hours.

The Bulova chronograph that belonged to Colonel David Scott – one of the 12 men, as of today, who have walked on the moon – is the only personal watch to have been worn on the moon, in 1971.

Back on earth, the astronaut carefully kept this watch, knowing its historic value better than anyone else. In 2015 he decided to sell it, and wrote a five-page letter of authentication: 'The Bulova Lunar EVA (Wrist) Chronograph and attached Velcro wrist strap...was worn by me on the lunar surface during the third EVA of Apollo 15, and then in lunar orbit and return to Earth...The primary use of the wrist chronograph on the surface of the Moon was to track...the elapsed time of consumables use (oxygen, water and battery) in the Portable Life Support System (PLSS) backpack...Our mission was to basically double the capabilities and requirements of previous missions, including especially the duration of EVAs outside the Lunar Module...Time is of the essence during human lunar expeditions – and exploration time on the surface is limited by the oxygen

THE ONLY PERSONAL WATCH EVER WORN ON THE MOON

and water (for cooling) we can carry in our backpacks...Knowledge of precise time remaining was essential...as a backup to the standard issued Omega chronograph, I carried and used a Bulova chronograph on the lunar surface.' A famous photograph shows Scott saluting the American flag on the moon; the Bulova can be made out on his left wrist. This was a source of great pride for the watchmaker's CEO, Omar Bradley, who had moved heaven and earth to try to outdo Omega when it came to American industrial patriotism. Several on-board instruments in the module used by the Apollo 11 mission were fitted with his company's Accutron mechanism, featuring a transistorized tuning fork and electronic oscillator; and in 1969 he scored a success when Buzz Aldrin and Neil Armstrong placed a Bulova chronograph on the Sea of Tranquillity to control the transmission of data back to earth – but he had to wait until the summer of 1971 for an astronaut to truly to fulfil his dream.

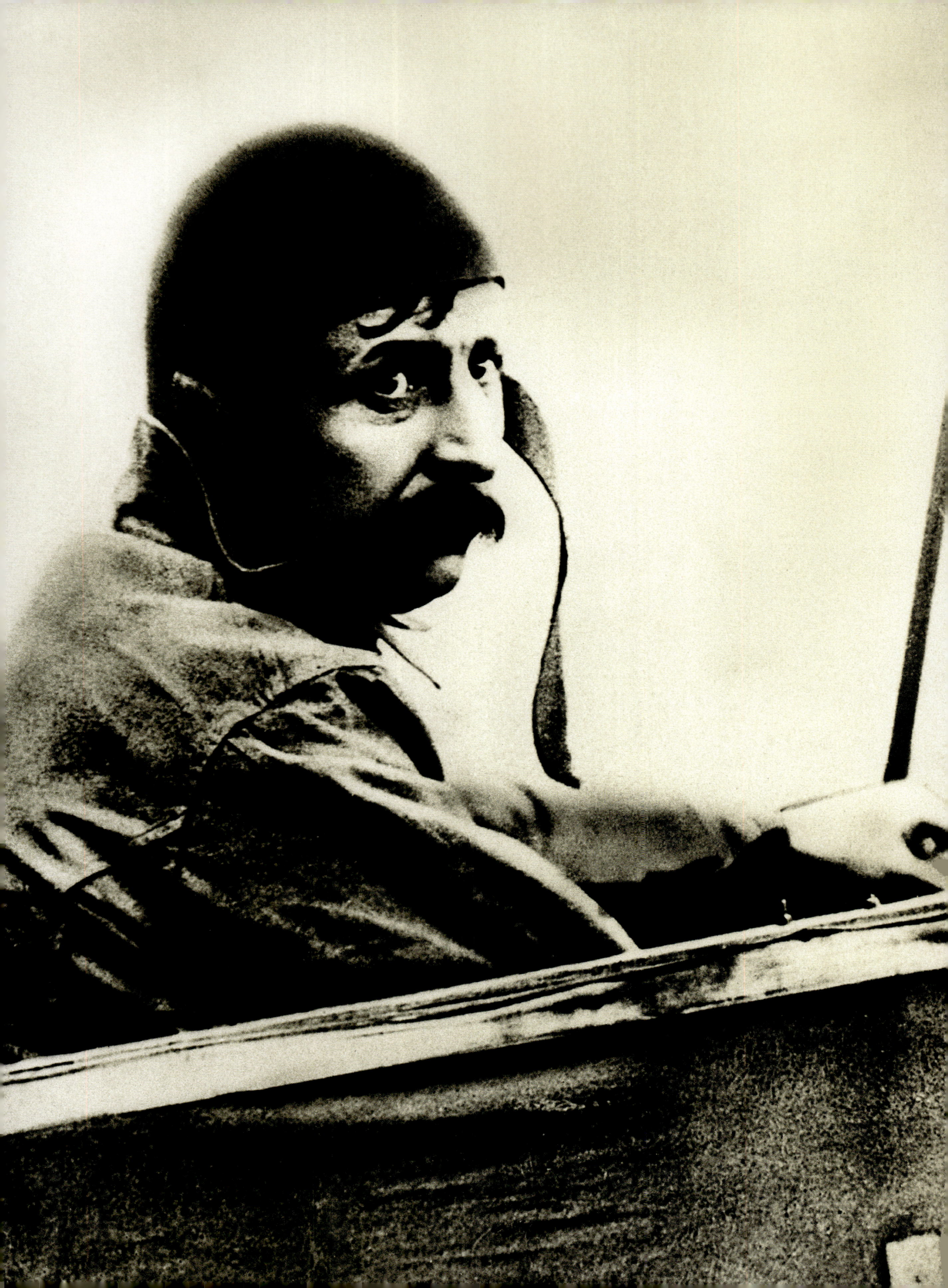

ZENITH AVIATOR'S WATCH 'LOUIS BLÉRIOT'

■

THE PIONEER OF AVIATION

Zenith began to make on-board instruments for civilian and military aircraft in the 1910s. By 1938–9 the famous Zenith Type 20 was fitted to the cockpit of almost all French aeroplanes.

On 25 July 1909 Louis Blériot crossed the Channel from Calais to Dover on board a monoplane, the *Blériot XI*, built to his design. He was the first to accomplish this feat, and pocketed the 25,000 francs cash prize offered by the British *Daily Mail* newspaper, which had put up the challenge.

That day, the aviator wore on his wrist a Zenith mechanical watch, which already had all the features of future pilot's watches: a black dial, luminous oversized Arabic numerals and hands to ensure perfect readability, a large ribbed crown that could be operated while wearing gloves, a triangular pointer to calculate time and a movement featuring a bi-metallic and anti-magnetic balance spring to prevent deviations in accuracy in flight. On 19 March 1912 Louis Blériot wrote of his Zenith: 'I am extremely satisfied with the Zenith watch, which I use regularly, and cannot recommend it highly enough to people in search of precision.'

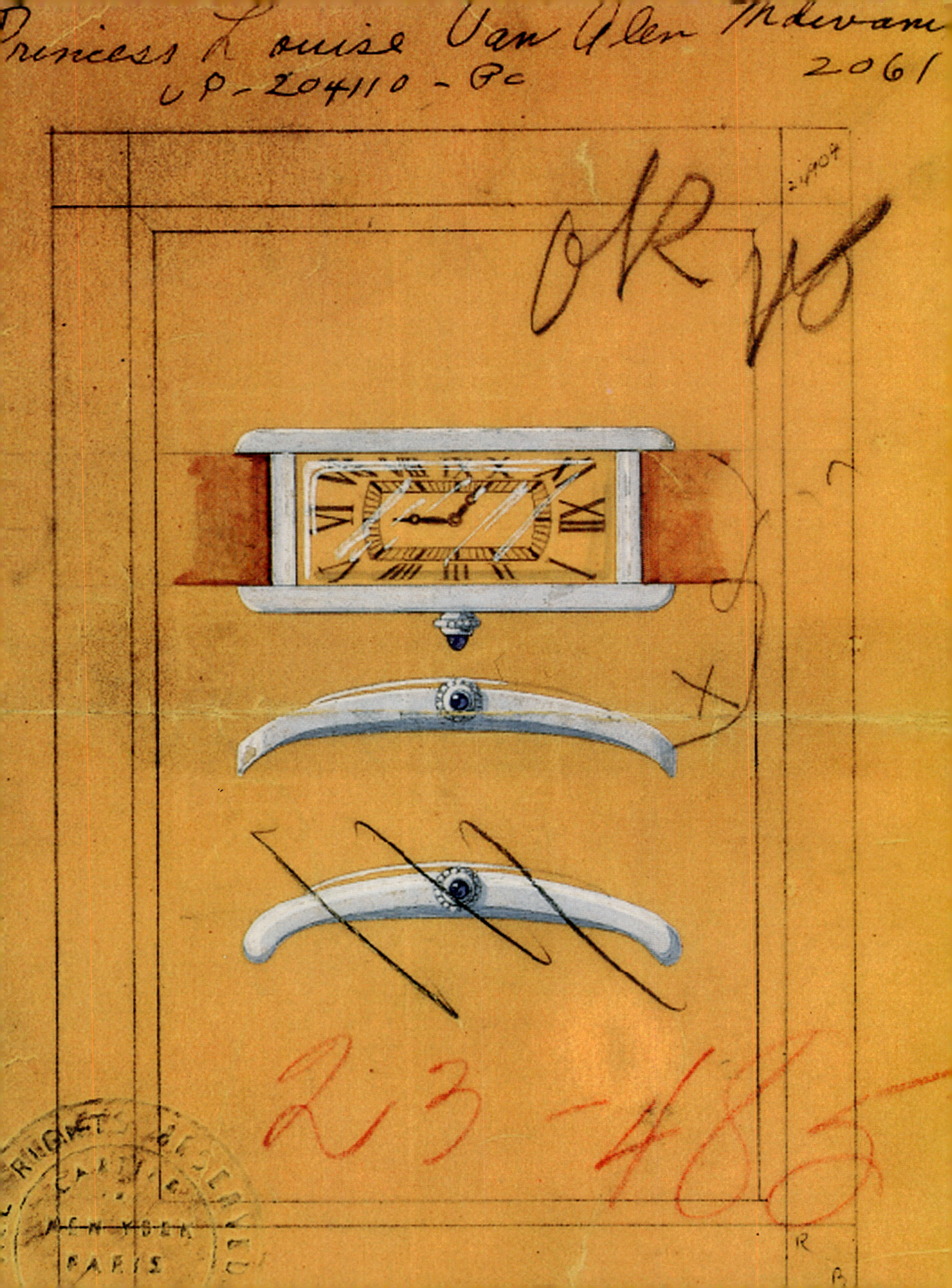
Princess Louise Van Alen Mdivani
2061
VP-204110-Pc
OK
23-485
PARIS

PORTFOLIO

■

THE 1,001 FACES OF CARTIER'S TANK

Probably no other watch made such an impression during its era as Cartier's Tank. Born of a powerful vision in 1917 and designed by Louis Cartier at the height of World War I, this watch defied the conventions of traditional watchmaking and sparked a stylistic revolution. The Tank rejected the circle in favour of the square, and established it in the world of watchmaking. It is thought that Louis Cartier himself approached the architecture of the watch with the idea of a battle tank seen from above – hence the watch's name. The lateral bars represent the caterpillar tracks and the casing the vehicle's hull. Production began in 1919, and thereafter the Tank underwent many transformations: classic, Chinese, solo, asymmetrical, American, oblique, French, rectangular, divan, tilting, crash, elongated, curved – without ever losing its appeal. Cartier's Tank was the first unisex watch in history: sought-after by men, adopted by women and idolized by stars. It was a watch of multiple models but at the same time unique – the absolute essence of French elegance.

Opposite: Drawing of a curved Tank design, Cartier New York, 1935, Cartier Archives, New York.

Andy Warhol was a great collector of Tank watches, especially the Louis Cartier model. The combination of the prince of pop art's rebellious attitude with the Tank's irreverent classicism worked perfectly. 'I don't wear a Tank watch to tell the time,' he explained, in typical provocative style, in 1973. 'Actually I never even wind it. I wear a Tank because it is the watch to wear!' Remove a watch's original function and retain only what makes it representative of a particular way of life – pure Warhol, in other words.

Above: Tank Louis Cartier wristwatch, Cartier Paris, 1944, Cartier Collection.

Following pages: Clark Gable and Jackie Kennedy.

The pianist, composer and conductor Duke Ellington was a keen collector of watches. An extremely rare 1946 Patek Philippe flyback chronograph that once belonged to him was sold for almost £1.2 million ($1.5 million) at Christie's in Geneva on 11 November 2013. Duke Ellington also loved Cartier's Tank watches, especially the models with windows, like this 1928 wristwatch, with a satin and polished gold case, which has a system for displaying the time via two windows. This watch's avant-garde style is matched by a rare and sought-after watchmaking complication: namely, the jumping hour.

Above: Tank watch with windows, Cartier Paris, 1928, Cartier Collection.

Following pages: Yves Montand and Rudolph Valentino.

OMEGA WRIST CHRONOGRAPH 'LAWRENCE OF ARABIA'

■

LAWRENCE OF ARABIA'S 'CUSTOMIZED' CHRONOGRAPH

On 19 November 2000, at an Antiquorum auction held in Geneva, the Omega Museum bought a historic piece with a mysterious past. For £65,900 ($84,600), the collection of the Biel-based brand acquired a wristwatch from 1915 with a single-button chronograph, which had belonged to the British archaeologist, spy and writer Thomas Edward Lawrence, better known as Lawrence of Arabia.

After painstaking study, Omega's historic division threw light on the origin of this uncommon watch. First, its calibre's serial number (4'428'513) proved that it had been ordered on 23 September 1915 by Louis Brandt & Frère, then Omega's agent for France and its colonies. After that, things got more complicated: 'The serial number 4'789'732 on the back refers to a standard 17-ligne (38.42mm) savonnette pocket watch, ordered on 17 September 1912 by the Liverpool dealer Joseph Sewill,' explains the brand's comprehensive reference book, *Omega: A Journey Through Time* (2007).

The 1915 Omega wristwatch with single-button chronograph that belonged to the British archaeologist, spy and writer Thomas Edward Lawrence, better known as Lawrence of Arabia. Note the bright enamel dial and the ribbed 'Louis XV' winding crown.

The experts' conclusion regarding the modified watch that had belonged to the author of *Seven Pillars Of Wisdom* (1926) was as follows: the inside of the back had been enlarged by hand, in such a way as to fit an 18-ligne (40.68mm) chronograph. The latter's dustproof cuvette had been removed in order to allow this replacement back, flatter than the original, to be secured when it was closed.

How did Lawrence of Arabia obtain this watch? Why was it modified in this way? These remain mysteries. We know only that its owner had it serviced on 18 April 1933: a guarantee slip delivered on that date to T E Shaw, one of his military pseudonyms, is proof of this.

A WATCH AS MYSTERIOUS AS ITS FAMOUS OWNER

The wrist chronograph has a silver casing, enamel dial and 'Louis XV' crown. Its hand-wound mechanical calibre is controlled by a gilded single button situated – unusually – at 6 o'clock. There are two engravings on the back of the chronograph: an 'A', underlined, and a wide arrow with three lines. The 'A' is a reference to aviation and the Royal Flying Corps (RFC), established in 1912 and replaced in April 1918 by the Royal Air Force (RAF). The arrow, known as a 'broad arrow', is a symbol still used on British Army equipment.

ROLEX DEEPSEA CHALLENGE

■

PLUNGED INTO DEEP WATERS

On 26 March 2012 the diving submersible *Deepsea Challenger*, piloted by the American film director James Cameron (known for *Terminator,* 1984, *Titanic,* 1997 and the *Avatar* series, 2009–present), descended to 10.908km (6.778 miles) in one of the deepest places in the earth's crust – the Mariana Trench in the Pacific Ocean. Securely fastened to an articulated arm of this civilian submersible, used for scientific work, was a Rolex diving watch. It had been developed, tested and made for the occasion, and it withstood the extreme pressure and functioned accurately in spite of it. This was the Deepsea Challenge.

In 1926 Rolex filed a patent for a screw-down bezel, caseback and winding crown, attached to a central case. This was part of the first ever waterproof wristwatch, named 'Oyster' because it could remain watertight under all circumstances. Ever since, the brand's supremacy in this field has been unchallenged. The experimental watch Rolex Deepsea Challenge is the ultimate proof of this; with its specific technical features, it is unique among its kind.

As robust as a submarine, it was designed to withstand a depth of 15km (9 miles) and is certified waterproof down to 12km (7.5 miles). At these extraordinary depths, the pressure on the glass is about 17 tonnes, and almost 23 tonnes on the back of the casing – some 40 tonnes in total.

Withstanding such pressures is a feat that brings to mind another prototype Rolex watch, dating from 1960 – the Deep Sea Special. This watch was on the bathyscaph *Trieste* when it made its historic dive, also in the Mariana Trench, down to 10.916km (6.8 miles), thanks to the visionary bravery of Swiss oceanographer Jacques Piccard and US Naval Lieutenant Don Walsh.

In fact, during his 2012 dive James Cameron took a 1960 Deep Sea Special with him in the submersible's cockpit. Afterwards, he explained that the watch had kept him company during his dive to the depths of the ocean and was, for him, a lucky charm.

From a technical perspective, the Rolex Deepsea Challenge (51.4mm in diameter and 28.5mm thick) is a reinforced version of the commercial Deepsea model, certified waterproof to a depth of 3.9km (2.4 miles). The Deepsea Challenge's monobloc Oyster casing is made from an ultra-strong alloy of 904L steel. Its structure is divided into three parts: an extremely strong nitrogen-alloyed stainless-steel ring; a sapphire glass in purest aluminium oxide, 14.3mm thick and slightly domed; and a 5.3mm screwed back in grade 5 titanium. This automatic diving watch is almost indestructible.

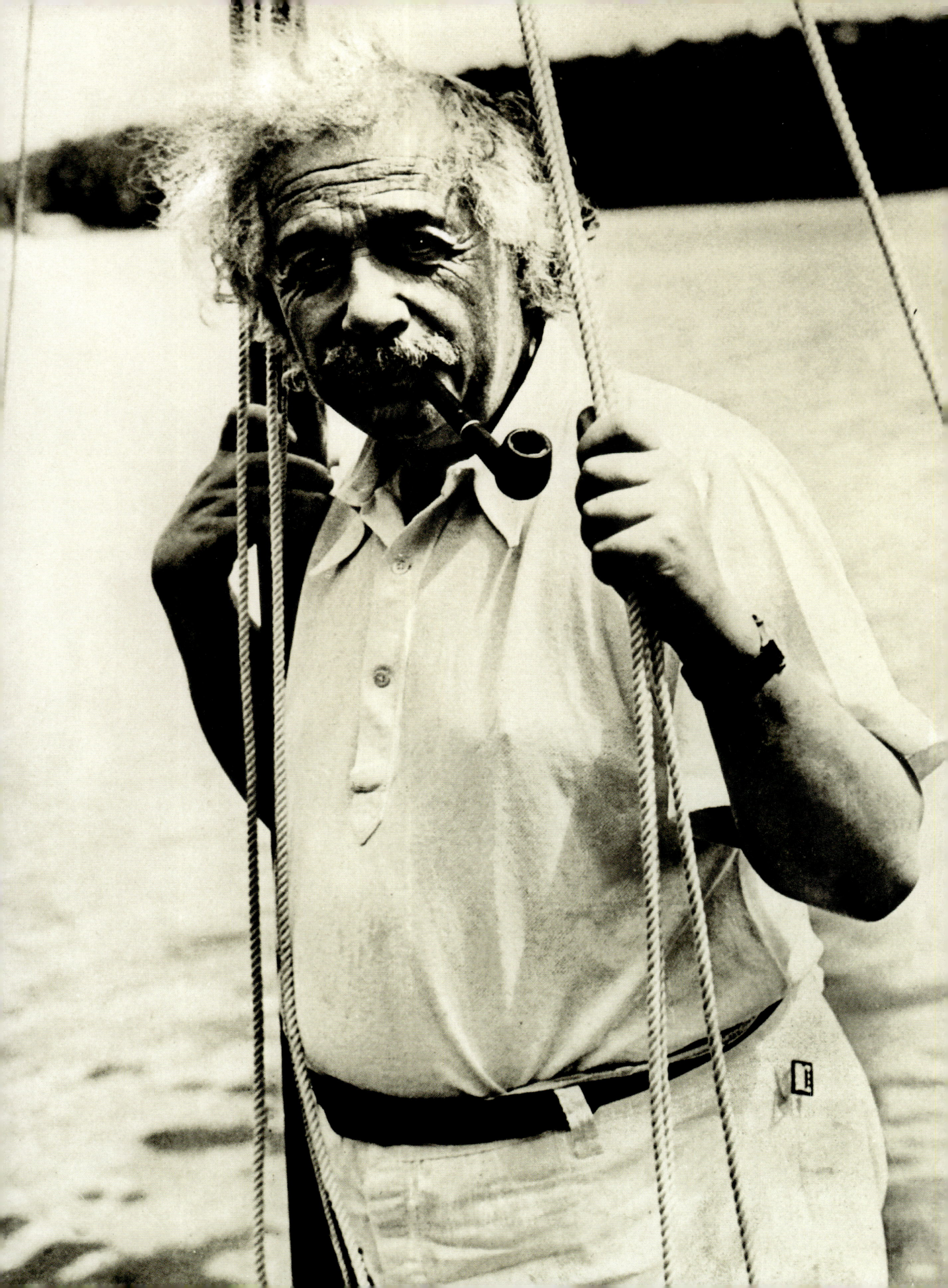

LONGINES ALBERT EINSTEIN

TIME, ACCORDING TO EINSTEIN

The epitome of aesthetic taste in the 1930s, Albert Einstein's yellow gold watch is, as of today, the most expensive Longines ever sold at auction. Sober and serious in style, it was the famous physicist's faithful companion for several years.

He revolutionized the concept of time. The father of the special theory of relativity (1905), Professor Albert Einstein was unquestionably one of the greatest scientists of the 20th century, if not the greatest in the history of humanity. Declared person of the century by *Time* magazine, Albert Einstein was given a Longines wristwatch on 16 February 1931 in Los Angeles. It has a Longines 10.86N mechanical calibre, a hand-wound 10-ligne movement – a classic of the time – and a barrel-shaped casing in 14-carat gold, just 25mm wide by 40mm long – minimalist dimensions by the standards of the day. The hands are very delicate; there is a railway minute track and numerals in a typeface typical of the 1930s. It is both simple and elegant.

The 1921 Nobel Laureate in Physics was presented with this watch by Edgar Magnin – known for being rabbi to the stars of Hollywood – at a gala reception held in honour of Einstein when he was 51. On the back is a three-line inscription: 'Prof. Albert Einstein / Los Angeles / Feb. 16. 1931'. It is said that he wore it daily; several photographs bear witness to this, including a very famous one that shows the scientist, stripped to the waist, on a sailing boat. He kept the watch until his death, at the age of 76.

A MINIMALIST WATCH AND A FAITHFUL DAILY COMPANION

On 16 October 2008 this understated watch, devoid of any complication – almost commonplace despite the personalized engraving on the back of its casing – was sold at auction by Antiquorum in New York. The buyer's identity remains a mystery; this piece of history was bought by an anonymous collector for $596,000 (£464,284) – 20 times its estimate. Never before had a Longines watch reached such heights.

Albert Einstein must have had a weakness for Longines watches; he also owned a Longines pocket watch from Saint-Imier, dating from 1943, which today is in the Bern Historical Museum.

AGASSIZ WATCH CO. & LOUIS COTTIER NO. 44497 'CHARLES DE GAULLE'

■

THE GENERAL'S 'VICTORY'

At the end of World War II, some citizens of Geneva got together to give an exceptional pocket watch to each of the commanders of the Allied forces: Winston Churchill, Harry S Truman, Joseph Stalin and Charles de Gaulle.

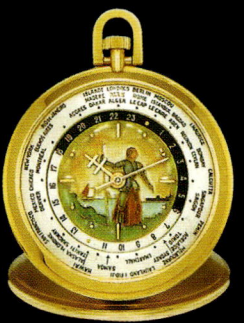

Each of these four 18-carat yellow gold timepieces featured the World Time function. They were designed in 1945 by the Agassiz Watch Co. in collaboration with the watchmaker Louis Cottier, inventor of the World Time function, and feature an enamel dial with personalized decoration by the famous Stern company and the enamellist Michel Deville.

While Winston Churchill's watch features St George slaying the dragon and a trident-shaped hour hand, the watch given to Charles de Gaulle shows Joan of Arc planting the Cross of Lorraine on the coast of France, with the hour hand stylized in the shape of a Cross of Lorraine, symbol of the liberation of France from Nazi Germany.

On the back is engraved a large 'V' for victory, as if inserted into a globe, beneath the inscription '1939 – Général Charles de Gaulle – 1945'. This watch was given to Charles de Gaulle in early 1946, accompanied by a letter: 'To General de Gaulle, president of the government of the French Republic, for the ardour of his faith and hope in the victorious French Resistance, the homage of the citizens of Geneva, with whom some French came together, in common admiration for the heroic battle begun on 18 June 1940. Geneva, Christmas 1945.'

At an Antiquorum auction held in the Hotel des Bergues in Geneva on 14 October 1990, this historic watch sold for £86,900 ($111,600).

INDEX

CREDITS

ACKNOW-LEDGEMENTS

I would like to thank all those who agreed to meet with me, help me and answer my (sometimes absurd, often silly and numerous) questions.

First and foremost I want to thank Aurel Bacs and Jean-Claude Biver, who generously agreed to collaborate in making this book by sharing their expertise and boundless passion for watchmaking. Esteemed authors of the prefaces, I thank you from the bottom of my heart.

My warm thanks also to Vincent Daveau for the pertinence of his remarks and the rigour of his 'technical' re-reading. Thank you, too, to Hervé Borne, Emmanuel Breguet, Thierry Castagna, Marine Lemonnier-Brennan and Justine Séchaud for their help and advice, as well as to Édouard Miquel, my sadly missed grandfather, without whom my love of watches would never have existed.

I am grateful to all the press and communications offices of the big auction houses, museums and watch manufacturers that had the good grace to make photographs and numerous technical and historical documents available to me; without these, this book would not have been possible. My heartfelt thanks to A. Lange & Söhne, Audemars Piguet, Bell & Ross, Blancpain, Breguet, Bulova, Cartier, Chanel, Chopard, Christophe Claret, Dior, François-Paul Journe, Franck Muller, Graff Diamonds, Greubel Forsey, Hamilton, Harry Winston, Hublot, Jacob & Co., Jaeger-LeCoultre, Leroy, Longines, Louis Moinet, Louis Vuitton, MB&F, Montblanc, Omega, Panerai, Parmigiani Fleurier, Patek Philippe, Philippe Tournaire, Piaget, Richard Mille, Roger Dubuis, Rolex, TAG Heuer, Tiffany & Co., Vacheron Constantin, Van Cleef & Arpels, and Zenith, as well as Antiquorum, Artcurial, Artsphere, Bacs & Russo, BlackDress, Christie's, DM Media, Dresscode, the photographer Étienne Delacrétaz, Hoda Roche Communication, L. A. Mayer Institute for Islamic Art, Laurence Phitoussi Communication, Liliane Fretté Communication, Magnapresse, Monaco Yacht Show, Musée Omega, Phillips, Publiquement Vôtre, Only Watch, RR Auction and Sotheby's.

Finally, thank you to Alice, Anne-Sophie and Félix for their support: to have a husband or father who is obsessed by the march of time, the colour of a dial, the rarity of a retrograde movement or the shape of a watch's hand is no laughing matter. This book is modestly dedicated to them, with all my love.

Paul Miquel